'It's a tough job, but it's a skill job. You have to maintain your instincts at all times. You can't let your mind wander ... You miss your family. But I handle it. I've been at it a long time. It's a job I have to do.'

—"Stretch," an apple-picker from Jamaica, who spends 10 weeks every year at a farm in Western New York.

Farm Hands

Hard work and hard lessons
from Western New York fields

Adapted with permission from *The Daily News* in Batavia, N.Y.

Printed by Hodgins Printing in Batavia, N.Y.

Fifth Printing, January 2011

Cover design by Brandon Capwell and Tom Rivers

Front cover photo by Angelo, a farmworker with Triple G Farms in Barre

Back cover photo (dairy farm) by Mark Gutman

Back cover photo (cabbage worker) by Nick Serrata

ISBN 978-0-9845656-0-3

Dedicated to Western New York farmworkers from near and far

Tom Rivers, center, is pictured with six apple pickers from Jamaica, part of a crew at Orchard Dale Fruit Farm in Point Breeze, N.Y. The workers include, front row, from left: Stretch, Clifford and Hall. Back row: Ferdinand, Shirley and Kereon.

- Table of Contents -

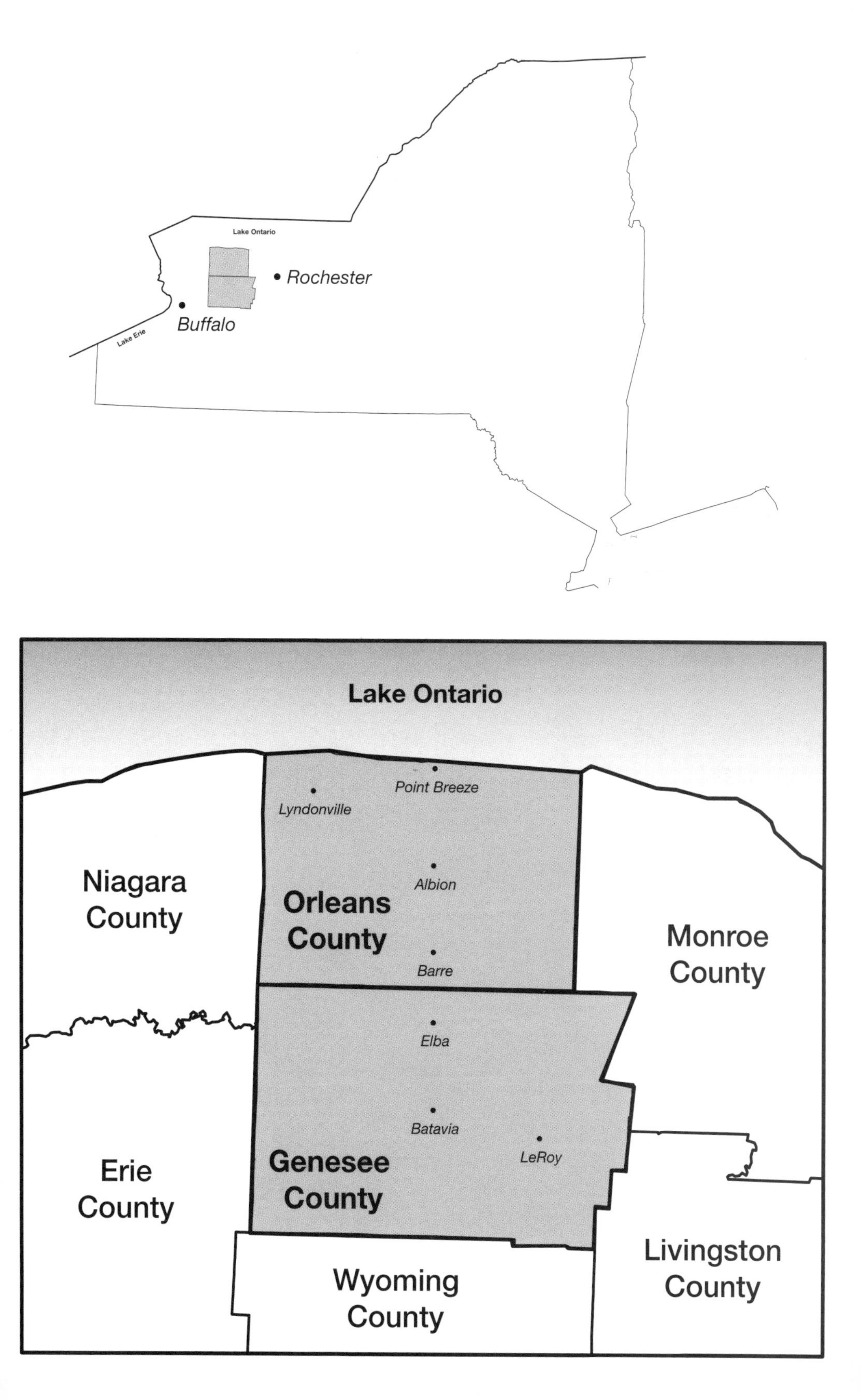
Lake Ontario
Rochester
Buffalo
Lake Erie
Lake Ontario
Point Breeze
Lyndonville
Niagara County
Albion
Orleans County
Monroe County
Barre
Elba
Batavia
LeRoy
Genesee County
Erie County
Livingston County
Wyoming County

Introduction:
Missing the big story

I didn't want *The Daily News* of Batavia, my employer since 1997, to do a *mea culpa* in 40 years, apologizing for a reporter who couldn't figure out how to cover one of the biggest stories in the community.

This is what I feared: "It has come to the editor's attention that the *Herald-Leader* neglected to cover the civil rights movement. We regret the omission."

The *Lexington Herald-Leader* made news in July 2004 when the Kentucky newspaper apologized for not covering the civil rights movement. The newspaper's reporters and editors were planning a 40th anniversary special, looking back at how the paper covered the black community in 1964, when residents marched in the streets, held sit-ins at restaurants and worked the back channels to gain more opportunities for jobs, education, housing and other rights.

Reporters discovered the newspaper had mostly ignored what was happening in the black community. The paper printed the above statement, acknowledging its previous failure.

It made me wonder whether *The Daily News*, a small daily newspaper in rural Western New York, was missing a big story, perhaps a sweeping issue that might not be so obvious. It didn't take me long to find a neglected, yet significant story.

I cover agriculture for *The Daily*, an important beat in a farming community. The local fruit, vegetable and dairy farms pump about $400 million into the local economy of three rural counties: Genesee, Orleans and Wyoming. Farmers had been telling me for several years that they struggled to find workers to milk cows and harvest crops. Many of the workers they did find lacked proper documents to be in the country. It made for a tenuous situation. Those workers, labeled "illegals" by many in the community, were vulnerable to deportation, leaving farmers, already operating with a lean workforce, without enough employees.

I had relayed farmers' concerns in many articles, and those stories elicited some strongly worded emails and letters to the editor from readers who insisted farmers were bringing in cheap labor from Mexico and rejecting local residents. I tried to update readers with many of the proposals for immigration reform and why farmers supported or opposed the plans.

But something was missing in those articles: the workers. They were rarely pictured, seldom quoted and their contributions were largely unappreciated. Farmers told me field work was physically grueling, that few local people could last in the jobs. But the public didn't seem to buy that. Many of the readers had worked on small farms as teens. They figured if they could do it, the current crop of locals should meet the challenge.

There are an estimated 5,000 migrant workers in Western New York who come each summer and fall to work on fruit and vegetable farms, with about 3,000 in Orleans, Genesee and western Monroe counties. In the past 10 years, more dairy farms have turned to reliable, hard-working Mexicans to milk cows, deliver calves and nurse sick animals. The dairies have undergone a transformation in Western New York, going from small operations mostly staffed by family members and a "hired man" to farms with 20 to 30 workers, with the owners serving as managers. Many of these farms now milk 1,000 cows or more.

In recent years farmers have tried to make the case to the community and Congress that agriculture's most pressing issue is a stable and legal workforce. All the big corn, vegetable and fruit prices don't mean anything without enough people to bring in the crop. Many of the dairymen and fruit and vegetable growers from WNY have visited Washington D.C., walking the halls of Congress twice a year, trying to press the cause. They have gone personally, hoping their direct pleas would help move the issue. They haven't left the work to professional lobbyists.

The people from Mexico weren't just making an impact at local farms. Some of the workers didn't return to Mexico. They made new lives in Albion, Medina, Holley, Elba and other communities covered by *The Daily News*. Some of the workers married, had children, purchased homes and opened businesses. I wrote about a few of the new Mexican businesses – a grocery store in Medina, a Mexican restaurant in Albion. I wrote about the farmworkers' soccer league in Hamlin. The family that owned the store in Medina, the Rosarios, invited me to their daughter's Quinceañera. They allowed me to write about the festive coming of age ceremony, when the Mexican culture celebrates a girl's 15th birthday, a milestone that signifies the girl is transitioning into adulthood and ready for more responsibilities.

I spent a few hours at Nancy Rosario's Quinceañera. *The Daily News*

Nancy Rosario poses with her aunt Beatriz and her uncle Pablo during a Quinceañera celebration for Nancy in July 2006. The event is a popular Mexican custom for girls who are 15 and officially becoming young ladies with more responsibilities. (Photo by Mark Gutman)

gave the story good play in July 2006, with many photographs in the paper. A few readers even wrote to thank me for the story. I thought we were doing better, covering an emerging cultural force in our community. I didn't think the paper would have to apologize for its lack of coverage.

In 2006, farmers became increasingly vocal about stepped-up immigration enforcement that was leaving farms short on help and families splintered. Batavia, besides being an agricultural hub, also hosts a 500-bed federal detention facility. Many farmworkers, often initially stopped for traffic violations, spend time at the detention facility because they cannot produce legal documents authorizing them to be in the United States.

Some of these workers married legal residents. Those workers might have children who were born here. When one member was deported, the families were split up. I reported about the "humanitarian crisis," how some members of the community, including a few religious leaders, were pressuring law enforcement to not aggressively enforce immigration laws. They didn't think people should be stopped in grocery stores, at church, the Laundromat or at a medical facility.

But the community continued to speak out against the Mexican farmworkers and the farms that hired them. I base that on an on-line survey by Tom Reynolds, an area congressman until he retired Dec. 31, 2008. During the 2006 election cycle, Reynolds asked constituents on his Web site if they supported efforts to overhaul the country's immigration laws, granting legal status to people who entered the country illegally. Reynolds said survey respondents were nearly unanimous in opposing efforts to legalize the estimated 12 million people in the country without proper documents. Reynolds cited the survey in explaining his support for stricter enforcement and stronger border security.

I was surprised by the lack of support for the farmworkers. I thought I had been giving people a steady dose of their value to agriculture, the biggest local economic force. But many people told me they couldn't approve of workers who entered the country illegally. There also was an assumption that local people would step in and work on farms, but that Mexicans worked for less pay. People thought farmers preferred Mexicans because they were cheaper employees.

In 2007 and 2008, farmers were increasingly frustrated with the inability to reform immigration laws. Many farmers told me the media wasn't helping to foster a meaningful discussion about the issues. Farmers said the media, particularly the conservative talk show commentators, were inflaming public passions against the workers, accusing them of overtaxing the social service system, bringing in diseases and living lawlessly.

In early 2008, with the dawn of a new election cycle, I was tired of merely relaying the congressman's comments, rehashing the farmers' stances about the need for immigration reform, and reporting about immigration "raids" at a local farm or store. I wanted to cover the issue in a different way, but I wasn't sure how to go deeper, to introduce readers to some of the workers and to try to show why so few locals wanted to work at farms.

Farmworkers play soccer in this photograph from July 2004. About 200 workers played soccer on Sundays at a park in Hamlin. The league has since moved to neighboring Brockport.

In March 2008, I had an idea, inspiration I couldn't shake no matter how hard I tried. That March I was one of about 20 people who attended a free screening in Orleans County for the film *American Harvest.* Angelo Mancuso, a film maker from Rochester, traveled to several states, interviewing produce brokers, major vegetable growers and farmworkers for the film. Mancuso has struggled to get a distributor for the film so it could be shown in theaters across the country. He goes to many colleges, screening the film, and tries to lead discussions about immigration and food, how farmworkers are crucial to a domestic food supply.

After watching the film at the Cornell Cooperative Extension in Orleans County, I wished there had been more footage in the trenches, out in the fields so you could really experience what the work was like. I thought Mancuso did a good job with his film, but for the next few days I wrestled with how the story could be better told. I thought a reporter without a film crew might be able to gain better access to the workers. I thought a reporter should try to actually do the jobs, let people know how hard it is, whether a regular American guy from the community could in fact do it.

Mancuso, with all the cameras, was no doubt rejected on many farms because the workers lacked legal standing to be in the country. If they were illegal, they wouldn't want a film crew capturing their faces. And farmers wouldn't want a bull's eye on them, drawing the attention of immigration and law enforcement officials.

I figured a hot shot reporter from a big paper, maybe *The New York Times*, should write the first-person accounts about farm work. That reporter could focus on the topic for six months or so, and not be burdened by the daily grind at a small-town paper. I didn't think *The Daily News* in Batavia, circulation about 13,000, was the right publication for this project.

But after several near-sleepness nights, I realized that *I* might be uniquely qualified for this project. I knew that if a reporter were to gain access to the farms, the reporter would need pre-existing relationships with the farmers, where a sense of trust was established. I had those relationships. I also knew the project couldn't be a witch-hunt about who is legal and who isn't. Maybe the farms and the workers would be open to me as a reporter if the focus was on what the jobs are like. In my reports, I would try to answer the question whether an average local guy could get through a day in the fields.

I happen to live in a very diverse agricultural community, with small and large fruit, vegetable and dairy farms. (New York is more than the soaring skyscrapers in New York City and the amazing waterfalls at Niagara. New York is the country's second-leading apple grower, the third-top dairy state and the fifth-ranked fresh-market vegetable producer — all requiring intensive manual labor.)

I envisioned trying different jobs at several local farms, and then telling readers how the days went, whether I was any good at the jobs, what kind of physical toll the work took on my body, and how I fit in with the workers from Mexico, Jamaica and Haiti.

There were two major obstacles facing me: about 30 extra pounds on my midsection and no farms eager to host me.

I had hoped to try jobs at maybe four or five farms. But once I found the first, then the second, and then the third farm willing to take a chance on this endeavor, many more doors opened, a dozen farm jobs in all. I even tried one job — apple picking — at three different farms, with workers from three different countries as my co-workers. I wanted to mix it up a little, to explore how there are varying methods to bringing in the crop.

And I shocked myself by shedding weight – 40 pounds by year's end. I'll even write about running a marathon, a feat I wouldn't have thought possible before working a long day on a farm. I discovered willpower I didn't know I had during this experience harvesting produce.

Before I move on and tell you the stories, I should be upfront about something. I met some inspiring folks in the fields. No surprise there. You'll meet many of them in the following pages. But you won't meet everyone. Some of the workers confided to me they were illegal, lacking the documents to be in the country. They didn't want to be pictured or have their names, even their made-up ones, in the newspaper. They didn't want to risk being sent home by drawing attention to themselves. They needed the money they were earning, typically about $10 an hour. Many put in 60 hours or more a week, with no complaints.

These workers, many in their late teens or early 20s, shared stories of long-term separation from their families in Mexico. Many of these workers intended to be in Western New York for maybe a year. But they hadn't been home in several years because of the tightened border. They worried they would be caught by law enforcement either going to Mexico or when they returned to the United States. Their families back home needed the money they were earning.

Maybe a hard-nosed reporter would have exposed them as criminals and alerted law enforcement of their whereabouts. But I didn't want to betray their trust or misrepresent my intentions with the farmers, who may or may not have known if their workers were legal. I told the farmers I wasn't doing an undercover exposé about which farmworkers were here illegally. My mission was to find out why so very few Americans are willing to work in the fields.

I met hundreds of farmworkers in 2008. Fabian was one of my favorites. He didn't have legal papers to be in the country so I kept him out of the newspaper. He was only 22 and had been away from his family, including younger sisters, for four years. He lived like a monk, spending little on himself. He told me he wanted to go home and visit his family, but he didn't dare try, fearing he would be kicked out of the U.S. if he was caught. He rarely left the farm. He said he would like to go to McDonald's, but even that felt off-limits.

He surprised me during one long shift together. I was tired and could barely keep going when he disappeared for about five minutes. He returned with an Intense Chocolate Mocha milk drink. That fired me up, all that caffeine and his kind gesture.

He knew me as Tomás. He told me a secret as I was leaving, after we spent a day alongside each other. He whispered, "My name is also Tomás."

Chapter 1
Finding a host farm and my former physique

Rich Gibson stands in front a tractor at Stein Farms. Gibson works part time at the farm. He installs furnaces for his full-time job. (Photo by Tom Rivers)

I've had some great ideas for the newspaper, but only a few of the projects have made it to print. It's difficult to do a series of stories or comprehensive report at a smaller newspaper. The reporters have many towns and municipalities to cover, stories to write about budgets, political races, zoning changes, festivals and churches celebrating 150th anniversaries. There seems to be an endless parade of community leaders and "characters" to profile. At *The Daily News* reporters are expected to churn out at least an article a day, and often I write two or three.

I'm not complaining. I like the job and even find the zoning discussions to be interesting. But I didn't see where I was going to find the extra time to work at a farm and report about the experiences. We were also entering a feverish political season with the 2008 presidential race wide open. We had added drama in the local political scene when long-time State Sen. Mary Lou Rath decided not to seek re-election. Tom

Reynolds, a local Republican congressman, also was ending his political career. I was the main political reporter for the paper, and 2008 promised high-stakes and high-interest politics. But I asked the editors and management to relieve me of that duty so I could focus more on the farm labor project. I told them the labor issue was the most pressing issue in the community, more so than the political races.

The editors agreed to reassign the political beat. But I still needed a plan. Soon the planting season would start, and I didn't have a place to go. I suspected that if a farm would take a chance on me, it would be Stein Farms in Le Roy. The Stein family is heavily involved in the community. Dale Stein is president of the Genesee County Farm Bureau. His sister-in-law Shelley Stein served eight years as town supervisor. Dale has gone to Washington, D.C., lobbying for immigration reform. I had a feeling the Steins might be open to me coming to work at their dairy farm for a few days.

It took a couple weeks before the Steins agreed. They were worried about me getting hurt. They didn't want their workers to feel uncomfortable or exploited. But ultimately, Dale believed this could be way to let the community know what farming is really like and to see what is being asked of the farmworkers. Shelley hoped the farm labor project would showcase careers in agriculture, enticing more local people to consider employment at farms, even people without a farm background.

In early April, Dale invited me for a tour of the farm, where employees milk 750 cows and care for another 750 heifers and "young stock." The farm is larger than most local dairies, but there are many bigger farms with more than 1,000 milking cows in Western New York. One farm near Batavia, Lamb Farms, milks about 4,000 cows. Dale showed me the Stein milking parlors and a series of interconnected barns with all the cows. Then he showed me a facility with the heifers, a calf barn for the newborns and another barn for animals that are about three to six months old — Holsteins that weigh 300 to 500 pounds.

Dale is in his early 50s. He is lanky and his knees are gimpy from years of walking up and down the steep steel steps on his tractor. The hours of walking on concrete floors haven't helped either. But the guy is tough, much stronger than I am. I realized that when he scooped a bucket full of a molasses-based feed for the three- to six-month-old calves. Dale easily filled the bucket as if the feed was water. I tried to scoop the feed and it felt like concrete. Dale did the chore with one hand. I needed both hands and could only get half a bucket filled. It dawned on me just how wimpy I had become.

I didn't have any visible veins in my arms, a sign of a guy who was too comfortable in life. I hadn't done a push-up in years, and occasionally, maybe once every two weeks, I might get in a slow jog. But since I married at 22, I had steadily added 2 to 5 pounds a year. I went from 150 to at least 190 pounds in about 12 years. (I stopped getting on the scale because I didn't want to know anymore.)

I needed to toughen up – and quick. I hadn't always been chubby with no muscle mass. I ran cross country for two years in high school. I also played soccer and was the leadoff batter and centerfielder for the varsity baseball team. I briefly broke the school push-up record as a freshman, only to have the school's top wrestler surpass me later in the day. I mention these exploits because at one point in my life I embraced punishing myself physically. I hoped I might be up for the fight again.

I started jogging in earnest the day after visiting Steins'. My knees didn't enjoy the 2-mile run. I took lots of walk breaks. My lungs also rebelled. But I stuck with it. For five or six days a week, I laced up my running shoes and ran along the Erie Canal Towpath. It took two weeks before I could go the distance without frequent rest stops. I don't think I lost much weight in those first few weeks, but I could tell my lungs and knees were adjusting to my new active life.

I returned to Steins' again for more observation. I was feeling overwhelmed by all the machinery, the barns, the new faces and all the animals. Dale and Shelley wanted me to take in everything — even to watch a vet do pregnancy checks on the cows. For my stint as a milker, I wouldn't be working with heifers or in the calf barn, so I didn't need to worry about most of what I was seeing. I still felt overmatched by the place and the barrage of activities, sights and smells.

But one of the workers eased my fears. Dale and Shelley suggested I ride in the manure spreader with Rich Gibson for an hour or two. Gibson, 30, has worked at the farm since he was 18, two years after moving from Jamaica. He is only part time at Steins' since becoming a certified HVAC technician. He works full time with Turnbull Heating in Batavia, installing furnaces.

Rich still works two days a week at Steins', where he said brothers Dale, Kenny and Ray are like fathers to him, and Shelley Stein is like a mother.

"They're good people – they treat you right," Rich told me. "If you work hard, they got your back."

Dale said Rich can run just about any machine and do almost every job at the farm. But Rich wasn't always so confident with his farm skills. He admitted he was scared when he saw all of the equipment and all of the cows. He milked five cows by hand in Jamaica for his grandfather's farm, an experience he thought qualified him for the job at Steins'.

The Steins initially bypassed Rich for the job in the milking parlor, giving it to a local American. But when that guy never showed, Ray

Stein went knocking on Rich's door, asking if he could start the job that day. Rich, who's now an American, told me the environment at the farm was totally out of his comfort zone. But he needed the work and told himself he would not quit. He tried to focus on mastering each of the milker's tasks, trying to improve with each day.

He learned to clean a cow's teats and attach the milker units. He ran the refrigeration system in the parlor and knew how to switch milk from one tank to another. He was good at moving the cows to the parlor and helping deliver calves.

He never called in sick, always arrived on time and sometimes stayed late, even working by himself on the night shift when someone didn't show up. Rich would become a valued employee, charged with training many of the new hires at the farm.

Rich is prone to philosophizing, even while driving a 5,300-gallon rig of cow poop on a soon-to-be planted field of corn.

"You don't know what's inside you," Rich told me. That sentence would become a rallying cry for me in the coming months, when I wondered how I would get through a day picking cucumbers, another day harvesting cabbage and other grueling farm jobs.

I still didn't have a planting job lined up. Dale suggested I try working in the muck at an onion farm. I called one onion grower and offered my services. I figured he would jump at the chance. I said I would work one day planting onions, and I'd do it for free. I just wanted the chance to write about the experience in the local newspaper. But this onion grower didn't want any publicity. "We try to keep a low profile," he told me.

He wished me luck, but told me he doubted any of the 10 onion farms in 5,500-acre mucklands, a highly fertile former swamp, would host me for a day. Drawing attention to a farm could attract immigration officers, and the farms didn't want to lose their workers.

I nearly gave up on the onions. But I thought one farm might bite. I'd known Guy Smith, owner of Triple G Farms, for a few years. I called him, and he told me he had a crew of legal workers through the H-2A program. There were no worries with the guys being arrested and detained by immigration. Guy said I was welcome to come out and give it a shot.

So on April 26, 2008 – my 34th birthday – I joined his crew of 13 workers, who had just arrived from Mexico, in planting onions.

Chapter 2

Planting onions requires fast fingers and strong back

Tom Rivers has muck smudged on his hands after finishing a shift planting onions, the first stop on his odyssey trying jobs at farms in 2008. (Photo by Angelo, a farmworker)

BARRE — I noticed the ads in local newspapers in January and February — many farms were looking for spring help in the fields. But farmers knew from past experience the ads wouldn't result in any good workers, people willing to brave the constant back-bending labor for about 10 hours a day.

The farms have to advertise the jobs, to make sure the local population gets a shot at the work. When too few local residents apply for the jobs, farms can then turn to a trusty supply of foreign employees, typically people from Mexico and Jamaica.

The farms seek help with the state Department of Labor to go through the federal H-2A temporary farmworker program. That's the route taken by Triple G Farms based in Barre. Triple G grows onions and potatoes on 400 acres of muck in Orleans and Genesee counties.

The farm paid to bus a crew of workers from Monterrey, Mexico, to Elba, where they started work April 24. They're guaranteed $9.70 an hour and at least 30 hours of pay a week. They also receive free housing and transportation.

The program is expensive for farms, but for Triple G, the cost is well worth it, says farm co-owner Guy Smith. His H-2A employees are legal and he doesn't need to worry they will be picked up by law enforcement, detained and deported, leaving him scrambling to get his crops planted and harvested.

Crates of onion transplants — "sets" — sit on a truck. The plants, grown for three months in Arizona , will finish growing in Western New York. (Photos by Tom Rivers)

Ahead of schedule

The warm and dry weather in April 2008 allows for an early start planting onions, and the crew of 13 workers has been laboring about 50 hours a week since their arrival. The farm is two weeks ahead of its usual schedule because of the warm weather and an eager crew, Guy tells me in late April, when I joined his crew for two days.

They seem at ease with the constant bending required to plant the onions. I don't see any wincing, any grabbing at their backs. They often chat with each other, laughing. I hear "Americano" mentioned a few times, with some giggles. Maybe it is my bright orange hat or my arms slathered in sunscreen, despite the overcast weather on my first day at the farm.

I manage to keep up with my new co-workers, somewhat. But I can't help but rub my back, shake my arms and legs loose, grit my teeth and pray for rain, especially on the second day, a five-hour stint — an unprecented amount of physical exertion for me. Four days earlier, my first try at muck work, the day's planting shut down after three hours because of rain.

Transplants and transplanter

Triple G plants four types of red onions and seven varieties of yellows. I have no clue what an onion plant looks like, or how they are planted. Guy and some of his workers give me a quick tutorial. We'll be working with Red Wing onions that already have been growing for about three months in Arizona. The transplants are about 6 to 8 inches long, with an onion bulb slightly emerging at one end. The plants resemble long, limp blades of grass, with roots like crazy hair that get tangled together. Just rip them apart and the roots spring back, no big deal, a farm employee named Alberto tells me.

The onion plants need to be pushed about an inch into the muck, spaced out every 4 inches. Triple G likes to use the transplants because they get bigger and better quality onions quicker than planting seeds in their fields. The transplants can be harvested in August. The farm also plants seeded onions and those will be ready later in the harvest season, in September and October.

The farm has a machine called a transplanter with eight seats. Eight of the 13 workers get a spot in a chair and feed the onions into clamps on a wheel. As a tractor pulls the transplanter, at a very slow 0.2 miles per hour — "creep speed" — the wheel moves and workers add the onions to the clamps. When the clamps hit the dirt, they open, setting the onions in the muck. Then two wheels on each side of the clamp push dirt around the plant, helping to stabilize it in the ground.

The machine makes a creaking sound, but mostly it's quiet in the fields. We can hear farmers in neighboring fields running their tractors. It's April 26, my birthday, and we can also hear small airplanes about 2 miles away.

The transplanter isn't perfect. Sometimes a plant doesn't stick in the ground or the worker can't untangle the plants in time or is reaching for a new batch of onions to feed to the clamps. When there are "skips" you need workers on foot to push onion plants into the ground or fill the missing spots. There seems to be a skip for every seven plants, with occasional breaks covering as many as 10 spots. That's why you need people on the ground, walking behind the transplanter.

A crew sits in a transplanter pulled by a tractor driven by Tom Vernaccini, operator for Triple G Farms. The transplanter has eight seats, while another group of workers walks behind the machine, pushing plants in the soil.

I am a walker the first day, hunched over looking for open

spots or misses in the black and moist dirt. Because of a noon storm, the wet soil isn't holding the plants too well. I feel useful, joining five others on the ground. We each have one or two rows. Some guys can easily handle two rows on their own. I trail in the back, tending to one row, sometimes two. When someone needs a new stock of onion plants, I try to cover more territory. We each get a wad of about 75 plants. There's a batch of them in baskets in front of the eight workers on the transplanter. The ground people might have a wad or two, and we need to replenish about every 20 minutes.

Keeping up

I consider this work an honor. I recently wrote about the muck's history, how the 7,000 acres of a former swamp was drained and cleared by Italian and Polish immigrants. They started small 5- to 15-acre farms on the muck in the 1920s and 1930s. About 1,500 acres of the original muck is no longer farmed due to erosion. The fertile soil has mostly vanished in one section.

My hands are caked in black soil. I think of the immigrants, imagining many of them working the same soil decades ago. With that thought, I try not to fall too far behind the Mexican crew. I know little

Triple G Farms uses the H-2A program to bring in a dozen legal workers. The farm has to provide them with transportation and housing, as well as a guaranteed $9.70 an hour.

Two workers from Mexico feed onion plants into clamps on a transplanter.

Spanish and they know little English, but I still want to send a message that I intend to work hard. They seem skeptical of me. Other local Americans have tried the jobs before and few have lasted more than two hours. Many give up after a half-hour. I want to at least put in a respectable showing, and not feed the "lazy American" stereotype among the Mexicans.

The ground crew likes the help, especially my first day, with so many skips needing onion plants on the wet soil. I'm an extra guy, easing a bit of their workload. I help the group to not fall too far behind the transplanter. When storm clouds again unleash rain a little after 3, we all sprint from the fields, seeking cover in a wooden storage barn, with flaking red paint. Triple G employee Tom Vernaccini, who drives the tractor pulling the transplanter, declares the work day was over. No one seems to mind. I am happy when one of the workers shows me where the water is so I could get a drink.

I walk back to my car when three or four of the workers make a point of coming over and shaking my hand. I feel affirmed.

Aches and pains

I am sore from all the stooping, but it really doesn't hit until Sunday, a day later. My hips ache. My lower back and shoulders are sore. I can't swing or pitch properly when playing wiffle ball with my nephews, part of my 34th birthday celebration – a day late.

Three hours in the field seems a little light though, especially when I want to "experience" the job. So I call Guy Smith and ask for another shot. Wednesday I am back at 7 in the morning. It is freezing. The Mexican workers are bundled up so you can barely see their faces.

We are all sniffling, and my nose drips to the ground. We go back and forth on the 300-yard field for about three hours when it's time for "café." An enterprising Latina woman from Batavia pulls up to the field in a pickup truck with an enclosed trailer. She serves coffee and snacks. Most of us have a quick drink and bite. The coffee is too hot for me so I set it inside my car. The Mexican workers quickly gulp theirs down.

Failure

We finish another pass on the field when one of the workers insists I take a seat in the transplanter. He grabs his back, pointing to me. He is concerned about me. The other workers have already rotated jobs. The ground crew seems to be the less coveted job, with the younger workers, those in their late teens and 20s, getting extra turns stooping in the dirt. Some of the workers appear to be well into their 40s. One woman, about age 20, also is part of the crew. She trails behind at times to talk on her cell phone.

You need to be quick to get the plant in the open clamp. And put the plant in right or else it won't stand in the dirt.

I assume the sitting job will be a breeze. Just set the plants in the clamps. The tractor is moving so slow after all. But I am immediately in trouble. I can't untangle the plants fast enough. I don't get the plants set in the clamps quite right. I maybe get one right for every four clamps, which are suddenly coming at me at a brisk pace. The guy next to me comments, "mui rapido" — very fast.

The guys on the ground complain. There are way too many skips. I ask for the tractor to stop and promptly give up my seat. I'm not sure I would get the hang of the job, and I don't want to kill the backs of the guys on the ground. I also don't want to incite a riot.

I am happy to return to my feet, gaining new vigor in pushing the plants into the muck. I also realize how good the crew is doing to only have an occasional skip. The guy who replaces me has about 30 in a row before he misses one.

Adios

After five hours, with a five-minute coffee and bathroom break, I feel I have hit my limit. Guy and I agreed beforehand for me to call it a day at lunch. "You don't want to kill yourself," he tells me when I boast I could probably do a full shift. I don't need too much convincing that a half-day gives me a good feel for the work.

By 12:30 I am done. We have covered about a mile with all the bending, following a step or two, back and forth across the field. I say goodbye — "adios" — and get a few handshakes. I try to tell some of the guys I will see them play soccer when their league starts in the summer.

The Mexican workers buddle up in the morning, when temperatures are barely above freezing in late April.

They play at a park in Brockport on Sundays, their day off.

I'd guess I planted 500 onions on April 26 and close to 1,000 on the second day.

I am clearly suffering when I leave the muck. I have to fold myself into the front seat of my Toyota Tercel. I stop by Triple G to chat with Guy. I can barely get out of the car.

I suspect if I kept at it, I could handle the eight-plus hours of bending, and maybe even learn how to place the onions in the transplanter. But it's definitely not unskilled or mindless work. You need stamina and hand-eye coordination. You need to be able to function on a team. Valuable skills, indeed, but in short supply for our local farms.

Chapter 3
Juggling many chores as a night-time milker

Tom Rivers follows a group of cows to the milking parlor. Most cows make the trip three times a day. (Photos by Mark Gutman)

LEROY — There's some kind of protest going on in the milking parlor at Stein Farms. A group of 18 cows have just been milked and they're not moving.

I say, "Come on, cows," a few times in a gentle voice, and still nothing but stares from the Holsteins. I try a louder voice: "Move, cows."

Nothing.

The cows usually walk the perimeter of the parlor after being milked and then walk to one of three barns. But on this mid-May evening at Steins', the cows aren't budging. My co-worker in the parlor, Steve Meiser, knows what's wrong. "They don't know you," he tells me. "You're new. They're a little leery."

Meiser climbs up the four stairs from the pit to the top of the parlor. "Come on, girls," he says to the cows. He gives one a pat on the neck. They start moving. He walks to the back of the line, tapping the cows on the back, telling them to move. He calls them "darlin'" and "honey."

In minutes, he has the cows out to the barn. He advises me to try a deeper voice with a little gruffness. He also suggests I ditch my bright-orange baseball cap. That sticks out too much. Try to blend into the environment, he says.

Meiser, 47, has been milking cows at Steins for 15 years. He wears a knitted gray cap and a gray sweater, typically sprinkled with brown cow manure. He has on green overalls and tall black rubber boots. His outfit is like camouflage in the barns, with the manure on the floors, sawdust in the stalls and black-and-white cows.

But Meiser does stick out with his constant "cow chatter." He tells them to move into the parlor — "Take your time, honey" — for being milked, and "There you go" when the group is done and released from the parlor gates. He can be heard shouting "may, may" or "Come on" while moving the cows back and forth from the barns.

Meiser showed me how to move 700 cows from the barns and to milk the 1,400-pound animals. I worked six days at Steins' in the night shift, the 3:30 p.m.-until-finished shift, which usually lasts until 2 or 3 a.m. I was there Wednesdays and Thursdays. Seldom has a job ever required so much of my body, brain and heart.

'You're going to get dirty'

Dale Stein, one of the farm's co-owners, meets me before my first shift starts May 7 and suggests I wear a wet suit — overalls that seem to repel water and manure. "You'll need it because you're going to get wet and you're going to get dirty."

Stein also gives me a pair of black rubber boots. I fit my sneakers inside the boots. I also get a "purse" that wraps around my waist and has two pockets, one for clean towels and one for dirty ones. (I'll talk about the towels later.)

I also get two forearm covers made of the same shiny stuff as the wet suit. Stein also points to the box of blue rubber gloves. He says to use as many as I want during the shift. The gloves will protect the skin on my hands from all the water contact or also fight bacteria from spreading to the cows.

Stein says I can leave early if I can't hack the work. He wishes me well and heads out of the milkhouse. I don't appreciate the suggestion that I may not last until the wee hours of the morning.

But Stein knows from experience. Many well-intentioned, determined, desperate-for-money people have quit within a few hours. They don't like the smell, the messiness and the demands of getting cows into the parlor, milked and back out to their stalls. About half of those cows will be milked twice in the shift.

Meiser estimates he has worked with 60 milkers through the years.

Most don't last a week on the job. He always worries when new people go for a break. Many times they don't return, leaving him alone to milk the entire herd.

"People don't like it, because to be honest with you, it's a hard job," he says.

Attack of 'the claw'

I open the door — "Employees Only" — that leads into the parlor. I've only been in a few of these before, just a quick tour when doing a dairy story for *The Daily News*.

There are already two rows of cows being milked, 18 on each side. The concrete floor has a few clumps of cow poop splattered around. I'm led to the pit, where two people milk the cows. I've never been down in one of these before. I'm told to grab some towels. Some are white and some light blue, the size of a wash cloth.

I get a crash-course in how to milk. First, after the cows are lined up, with their back ends facing the milkers, the cows need to be dipped. There are two wands with hoses, one at each end of the pit, that shoot iodine foam onto the cows' teats. This kills bacteria and loosens any manure on the four teats.

You need to extend your arm, line up the wand on each of the teats, and release the orange foam. It makes a swirlie sound.

This is apparently the easiest job in the milking process, although it can tire the arms. It also takes hand-eye coordination to line up the end of the wand with the teats, especially those off to the side of an udder.

When the cows are all dipped, the teats need to be wiped and then "stripped." I'm advised to take out a towel and wipe each teat to clean and dry. Then you "strip" them by making a couple pulls down to squirt milk. It takes several tries before I can squirt milk from the teats. You need to lightly squeeze the teat, and make a quick downward motion.

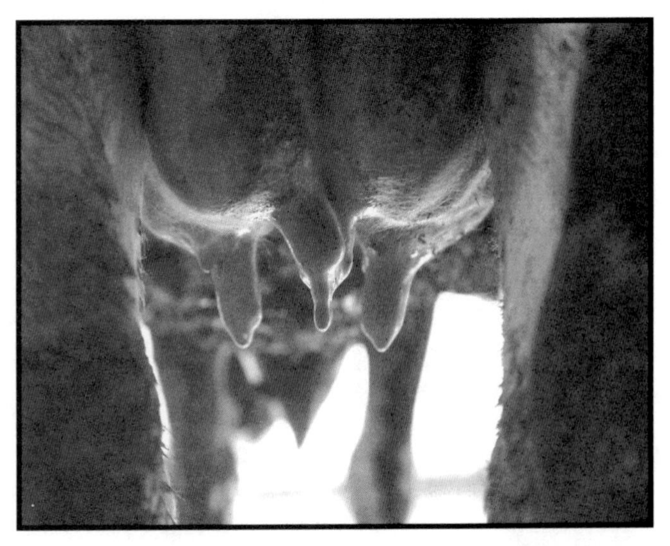

Iodine foam covers the teats on this udder. The iodine fights bacteria.

The milkers need to watch the

squirted milk to see if there are any white flakes. If there are, the cows may have mastitis, an udder infection. A few of the cows have it, and the milkers just squirt the milk until the clumps stop showing. If the milk is clear, instead of white, that's a sign of a more-serious case of mastitis.

The infection may just be in one teat and that teat doesn't get milked. The cow's number is added to a board in the milkhouse. The farm's herdsman will check the cow the next day, giving it medicine to fight the infection. The cow likely will spend a few days in the hospital barn for treatment.

After the teats are dipped, wiped and stripped, It's time to attach "the claw," a vacuum-like machine with four arms called inflations. The claw, officially known as a milker-unit, mimics a calf suckle. The milker needs to get one inflation on each teat.

It's a tricky task. The experienced milkers can quickly get two inflations on the two front teats near the cow's tummy and then fasten the other two on the teats near the back legs. The pros do this in about five seconds, making two movements, lunging in to do the front two and then leaning backward to do the other two. Then they move to the next cow in line.

I try to follow the milkers' example and quickly get attacked by an inflation. It latches onto my shirt sleeve. I shake it loose and another inflation starts inhaling the front of my shirt. It takes a few tries but I figure out how to hold the two back inflations together without having them sic me.

I quickly give up trying to do two inflations at once, like the parlor pros. I tell myself to focus on mastering the fundamentals before getting too fancy. I will be pleased to do one inflation at a time.

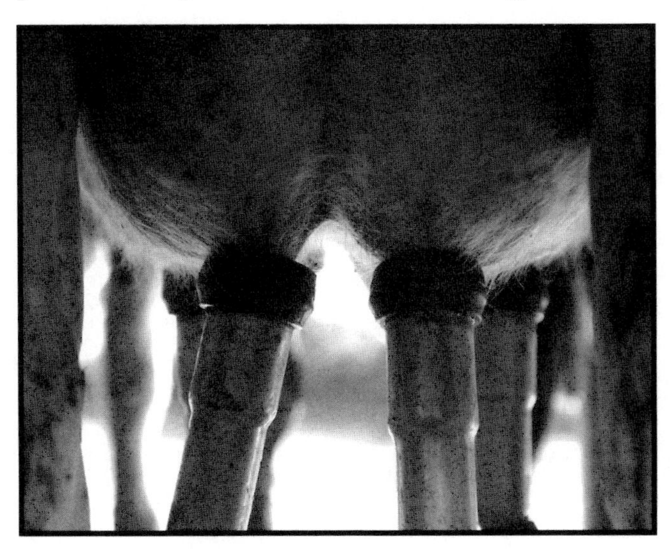

The inflations suck the milk out of the udder.

While holding three other pulsating inflations , I lean in to get one inflation on a teat. The inflation seems to jump out of my hand and latches itself on the teat. I can feel the milk being pulled out and run through a tiny hose. I do the other teat — not so bad. Then I go to work on the two teats near the back legs. The inflations quickly go on, without incident.

I'm done with the first cow and I wonder how the milker behind me is faring. He is no longer near me. During my scramble with the inflations, the other milker has hooked up all 18 of the cows on the other side.

He has grabbed the wand for a final iodine dip after the milking. That gives the cows added protection against bacteria while the teat ends are still open. They will soon close following the milking.

It takes a couple of hours of earnest effort, with my neck aching from nervous tension, and I think I can get the four inflations on a cow without being attacked by the claw. But just when I think I have things under control, an inflation sucks in the tip of the rubber glove on my hand. It makes a dreadful noise, like the whoopee cushion.

I get more tips on technique from the other milkers. I slowly get better with the inflations . It takes a few hours and I "improve" to getting the claw on four or five cows in the time it takes the "pros" to have 18 ready. I abandon any hopes of keeping up. I just want to do a decent job, even if it's at a slow pace.

'Crazy' udders

If all the udders were similar, with the same teat arrangement, the job wouldn't be so tough. But some of the cows only have three functioning teats, and one of those may be on the side of the udder. You have to reach right in front of a back leg, and there can be kicks, jumps or general uncooperativeness from the cows. They don't like their legs getting touched, it seems.

Steve Meiser washes the concrete floor in the milking parlor, one of many chores with the job.

Steve warns me he's taken a few kicks to the face through the years. So be careful, he advises.

I spend an inordinate amount of time on one of the three-teated cows before Steve says this particular model only has two working teats. One of the teats is closed,

Tom Rivers, front left, and Steve Meiser milk the cows at Stein Farms. There are 18 cows on each side. You might want to watch your step.

like a stub, with no milk coming out. The inflations not needed are plugged so the vacuum doesn't suck a lot of air into the system.

I discover a new appreciation for some of the 4-H blue ribbon winners I've written about over the years, those with the top cows in dairy shows. I remember some of these kids telling me why their cows were so good. They always mentioned udder formation, with the four teats set up in a square, so they would be easy to milk. I didn't appreciate that until encountering some cows with teats in wild angles.

Steve makes a point of looking for the three-teated cows, to spare me the struggle.

We have a few cows come in with gigantic udders — bigger than beach balls — leaving only a few inches from the ground and their teats. Just getting the claw under these cows is a challenge. Steve somehow gets the four inflations hooked to the cow. It takes a few more seconds than the usual cow.

"There's some crazy ones in here," he tells me.

The no-look technique

I gain some confidence by the second day in the parlor. I can't do the two- inflations -at-once trick, but I learn the no-look technique. Instead of squatting down, looking for a faraway teat, I just start reaching

This cow wonders about the stranger in the pit of the milking parlor. The animals are creatures habit and they notice anything out of the ordinary.

around, not seeing what I'm doing. I find a teat and direct the inflation. It latches on.

The no-look is quicker and much easier on the back and knees. I'm also less vulnerable to a kick in the face. But it means you may be pressing yourself against the back of some dirty cow legs, smeared in manure.

"You're not afraid to get dirty," Steve tells me, seeming impressed.

He observes my new no-look skills, and tells me I'm picking the job up pretty quickly. He thinks I could be a capable milker in a week. I'm still pretty slow.

"You can just tell with some people," he tells me. "Some people you know they'll never get it. But I think you can do this. I'll put in a good word for you."

I remind Steve about my newspaper project, that I'm just seeing if I can do the jobs, and trying to write about farm work.

Dancers

There are two types of cows that aren't so easy to work with: those with tails and the first-calf heifers.

Maybe 10 cows out of the 700 have tails that swish like windshield wipers. The fat tails nearly disrupt everything, blocking efforts to dip, wipe, strip and get the claw on the teats. I learn after a few days to just grab the tail with one hand or move it out of the way with my shoulder so I can get the teats cleaned, wiped and latched on.

But on more than one occasion, a tailed cow happens to pee just when I'm ready to start wiping and stripping, and she wags her tail, sending urine everywhere, including on my face, glasses and shirt.

The tails are usually removed when the cows are babies, but a few are missed for "hygiene reasons," I'm told. Thankfully there's only an occasional tail in the way.

There are far more first-calf heifers, the rookies in the milking parlor. Most of the veteran cows walk down the parlor, get their spot in the gates and slowly back up, without putting up a fuss. They don't even flinch with the dipping, wiping, stripping and latching. They almost milk themselves. I like these cows a lot.

But the first-calf heifers are like wild bulls. These cows have just had their first babies and these new milking mothers seem lost and shocked by everything in the parlor. When the iodine foam is shot on them, the cows jump, kicking their legs. Touch their udders, the cows jump. Try to fasten an inflation, and they jump, often knocking the claw off so you have to try again.

"They're dancers," Steve says. "They'll calm down once they're used to it. But be careful."

He would get kicked twice in a sore thumb that day, and he didn't call the dancing cows "darlin's" after that.

Incoming

I get hit with pee from the tailed cows on several occasions. Splattered cow manure on my boots, pants and shirt also becomes routine. I avoid a direct hit until the fourth day. I was wiping a cow, when the poop came crashing down. It just misses my head, but covers my left arm.

I must have shown my disgust because Steve notices the commotion, and the poop on my shirt and arm. "That's what we call an incoming," he says with a laugh.

Other workers have received direct hits down their backs, and needed to take a shower — immediately, Steve says.

When dumped on with poop, Steve suggests not spraying it with one of the water hoses. That would make an awful mess, he says. Instead, just wipe it off with some of the towels. I heed his advice. Then I give my arm a light hosing.

I don't mind the manure smell at all. I've been through the diaper stages with three kids. Poop doesn't bother me. The key I've learned is to make a quick attack and don't let the stuff linger. If you do, the smell is far more pungent.

I also didn't mind the poop splatters on my clothes, arms and occasionally face. Hey, it happens.

More than milking

Milkers spend most of their time — well, milking the cows — but there are many other duties, especially for the night crew, who are the only ones on the farm after about 7 p.m.

The first task is getting the cows to the parlor. They have to be retrieved from their stalls, which can be 500 feet away or more. Just opening the gates is enough to get most of the cows going and on their way to the parlor.

But some of the cows won't budge, at least for me. Steve just walks through the alleys, yelling "Come on, girls" and almost all of the cows get in motion. A few are lying down, and he either nudges them with his knee or gives them a firm pat on the back.

Moving the cows is much easier with two people. Many of the barns have several alleys and I notice a few cows will head down a different lane once they're "pushed." With only one person on duty, there's open alleys, and you can chase a cow in circles. So it's a good thing I'm there to head off some of the open lanes. Just standing in the way sent a message to a few adventuresome cows that they needed to follow the pack to the parlor.

I can't imagine one person trying to get 160 cows in a group up and moving in a coordinated fashion. Steve says it's not easy, especially with the sometimes clueless first-calf heifers.

When the group makes its way to parlor, the cows are in a holding area of sorts. I press a button that lowers a door or a gate panel, as it is properly known. The milker in the parlor then hits another button, which makes a loud ring and moves the door closer to the parlor, ushering the cows with it. That funnels the first group of 36 cows into the parlor, with 18 on each side.

It takes a good 15 to 20 minutes to do each side. When we're done milking a group of 18, a gate is raised and cows head out, making room for the next 18. But the new group of 18 cows doesn't just walk in. You need to ring the bell or say "Come on, cows" or "Move, cows." I develop some gruffness in my voice as the days wear on and the cows seem to notice I mean business.

I even ditch the orange hat one day, trying a dark green one from the tractor manufacturer Claas. But the hat is too tight, and I don't want to rub off any precious remaining red hair, especially in the frontal lobes. So I'm back to the orange hat, the only one that fits me, for my remaining farm days. The cows will just have to adapt.

Mission: poop removal

I notice on my first day in the parlor that one cow slips on a pile of poop and comes crashing on her stomach. It makes a dreadful noise, like a refrigerator falling from the sky and smashing onto the pavement.

But the cow gets right back up and seems unaffected. These animals have incredible determination. I would see a few Holsteins with a bad foot, either a painful wart or a soreness from hoof trimming, that left them limping. But these ladies still would make the trek three times a day to the parlor. They earned my admiration.

Tom Rivers directs liquid manure to a drain in the corner of the milking parlor. The poop can be a hazard to the cows because of its slipperiness.

After seeing the cow slip in the parlor, I made poop cleanup an obsession. If there was time in between groups, I grabbed a shovel and cleared the manure from the floor, directing it to one of the drains. I wouldn't see another cow fall down in the parlor again. But I saw one slip coming into the parlor on my second day. There's an incline requiring the cows to walk up. It didn't seem to be getting shoveled so I started doing that.

Steve would see me pounce on some of the poop piles on my second day, quickly clearing the potential pitfalls out of the way. "You're hired," he yells.

Later that night he would insist I try the Skid Steer, which looks like a fork lift except with a plow-like scraper on front. When the cows are led out of the barns to the milking parlor, one of the milkers needs to clear the barn alleys, directing the manure to a pit. It's cleaner for the cows, and no doubt makes it easier for them to walk.

I'm very tentative with this machine. I've never used one before. It doesn't have a circular steering wheel. Instead you steer it and direct the plow with joy-stick controls for each hand. I can't drive the thing in a straight line and nearly clip one of the stalls in my slow drive down an alley.

I suggest Steve run the Skid Steer. I don't think Dale, the boss, would be pleased if I broke something. I think he assumed I would be confined to the milk house.

The following week I return and discover the Skid Steer duties have been reduced in half. I'm told it's not to keep me from the machine, but instead to reduce diesel consumption. The fuel is costing $5 a gallon.

Many other chores

With each day, I try to add new duties, take on a new responsibility among the 20 different things these night milkers must do. By the second week, I'm doing loads of laundry every two hours or so. We go through probably 1,200 towels a shift. They all need to be cleaned and then dried.

The laundry machines are easy to run. Just hit the "start" button. I can handle that. I did manage to dump about half a tote of towels coming down the stairs into the pit. There was a hanger for the iodine wand that clipped the tote, sending the towels asunder.

I also try to be more active in rounding up the cows because it's such a pain for one person to get them all up to the parlor and then back to their stalls.

I learn to run the power-washer when it's cleanup time. I give the walls a thorough soaking with the strong blasts of water from the power-washer. It's a very satisfying chore.

But I'm not there enough days to learn to run "the system," where the hoses are all flushed and sanitized with cold and then hot water. The milkers redirect pipes on the top of milk tanks when one tank is full and the new batches of milk need to go to another tank. They do other stuff with the "the system," the series of tanks, pipes, heating and cooling tanks, and other stuff I don't quite understand.

They also change a filter every 8 hours and Steve is pleased with one that is nearly flawless, with a couple bits of dirt or manure and a trace of mastitis. "You're hired," Steve again declares.

The night milkers also have to check on the baby calves every couple of hours. Newborn calves need to be fed colostrum, the nutrient-rich first milk from a mother. The colostrum is in a frozen bag that needs to be thawed and then placed in a bottle. The baby calves also are moved away from the adult cows and into a separate pen with other babies that weigh about 75 pounds. The night milkers get to move the babies, which aren't steady on their feet yet.

There's an endless stream of laundry to keep up with because of all the teat cleaning. The night-time crew goes through at least 1,000 towels per shift.

The night milkers also need to be ready to help with any births. Sometimes the babies come out on their own, not needing any help. Other times the mother needs assistance, with a chain looped around the baby's front feet.

"It's pretty neat

Tom Rivers tries to move a group of cows to the milking parlor. There are always a few stubborn ones.

when they come out," Steve says. "You see them looking around. They're new in the world. I've done it hundreds of times and that's not an exaggeration."

A new career?

Steve wonders if it ever came to it, if I needed a job, would I consider working as a milker?

I tell him I have a ways to go to learn the job well. I don't know how to run "the system," with all the mechanical gadgetry. I also feel inadequate on the Skid Steer. I'd probably have a stress attack if I had to deliver a baby calf by myself, especially if it was a difficult birth. And I haven't been expected to carry the full load of an experienced milker in the parlor. I was the weak link in the chain with the other milkers. My colleagues seemed happy I showed up and didn't bail out. Anything I did was a plus.

Steve says I would develop all the skills in time, maybe quicker than I think. I like his encouragement.

Back to his original question: I tell him I would do whatever work I could do if I was suddenly jobless. The milking job certainly wouldn't be a last option. Stein Farms pays most of its milkers $10 to $13 an hour, with health insurance, vacation and holiday pay, and the farm contributes to a 401(k) retirement plan for the employees. The farm also offers life insurance and disability benefits. That's not so bad.

Plus you can rack up the hours, working 60 hours or more in a week if you're good at the job and willing to work. Not bad at all. In fact, for many people, that would be pretty darn good.

"Well, I'll put in a good word for you," Steve tells me again.

Chapter 4

Seeking a sweet reward: The hunt for cherries

Cherry pickers need a 16-foot-high ladder to reach the top branches. (Photo by Tom Rivers)

ALBION — It was a sight that put fear in my heart: a long steel ladder that seemed to reach to the heavens.

I thought picking cherries would keep me on my feet, just be a little stroll in the orchard. I almost didn't pursue the task because I figured it would be too easy. I want to try hard farm jobs after all, not stuff kids and grandmas can do U-Pick style.

But the ladder threw a big wrinkle in this assignment. I try to avoid ladders at all costs. At home I have a gutter choked with wiffle balls atop a one-story porch. I'm not in a hurry to retrieve them.

A cherry picker can't avoid work on a ladder. Probably 75 percent of the cherries are beyond what I can reach on my tiptoes. I don't like this harsh reality.

"I'm not really a ladder guy," I sheepishly confess Friday morning to the farm manager at Watt Farms on Route 98.

But Jesús "Chuy" Vallejo says I will soon get over my fears. He shoots up the ladder himself, comments on the view — "You can see everything up here" — even shakes the top of ladder on the tree to show it's securely lodged between some branches.

"There's nothing to be afraid of," he tells me, and then quickly climbs down what seems like the wall of a skyscraper.

I ask him if anyone ever falls off the ladder. "Sometimes," he says. He doesn't seem too concerned about me. He tells me I'll get used to the ladder. Just give it some time, he says.

Picking pointers

Chuy, pronounced "Chewy," hands me a wooden harvest basket that holds a half-bushel, about 20 pounds. He shows me the technique of picking cherries. First find the deep red ones, make sure they don't have "splits" on the bottoms, and give the stems a little twist near the branch. I try it and I manage to snap off a branch with the cherry.

Chuy lets out a sigh.

"That wasn't good?" I ask.

"No," he says, explaining that I just left an open wound on the tree that could lead to infection.

Chuy departs, urging me to just pick the ripe cherries, and show some care for the branches. Don't yank the cherries, he says.

Ground attack

Not at all eager about climbing the ladder, I opt to start on the ground, where there are many cherries to pick. I take out more tiny branches. They come off way too easy. I try to use both hands, holding the branch so it doesn't come with the cherries. My face is full of leaves. I wish I had worn my glasses instead of contact lenses.

There aren't too many cherries with split bottoms. In general they seem to be in superb shape. A few aren't ripe yet. I'm not sure where the cutoff is, for what is ripe enough to pick. I grab one that is a lighter red, and walk a few trees away to one of the four Hispanic workers picking. He says to leave them alone. They'll be ready in a few days.

I head back to my spot. There's a *Daily News* photographer with me for about 20 minutes but he has to leave by 8 a.m. I gingerly go up the ladder for a few photos, not at all comfortable. I'm slowly creeping up the ladder when my right leg starts shaking like mad. Nick, the photographer, seems to enjoy my misery.

I end that escapade after five minutes, not really doing much picking, focusing instead on not toppling to the ground.

Once Nick leaves, I'm pretty much alone, having six or seven trees to myself. One has a water jug propped on it. Some of the workers swing by every half hour or so to unload some cherries in lugs, hardy plastic totes that hold 32 pounds of cherries. The workers also stop by for a drink of water. I get a few waves.

Facing my fears

After a half hour of hitting all the low ones on one of the trees, I reckon I need to buck up, be a man, and work from that ladder. A few steps up and I have new batches of cherries to grab. If they're right near the ladder, it's not too bad. But many require a lean from one side of the ladder, with my arms stretched to the max. I feel a cherry, give it a pull, and off goes another branch. Another wound has been exposed.

Tom Rivers battles a face full of leaves while hunting for cherries. (Photos by Nick Serrata)

I gather as many cherries as I can within arm's length. But after 45 minutes of slow movement up and down the ladder, it's time to move the ladder to a new spot. Chuy had it propped right in the center of the tree, where all the main branches intersect at the trunk. The branches spin out in wild directions. The tree resembles Medusa's hair.

I try to move the ladder, which seems to be 25 feet high, but the thing is wedged between some branches. It just won't budge. So I move on to more ground work at the neighboring tree. I fill my basket, unload the cherries in a lug, and get psyched up to get the ladder freed from the tree.

For some reason I can't explain, maybe it's just the excessive quiet time alone, I adopt "Get 'er done" as my mantra for the day. I mutter the phrase and grab the end of the ladder, giving it a fierce tug. The ladder is unleashed and comes crashing to the ground with me beside it. Some leaves and branches also come floating down.

I'm thankful nobody is filming me in action. This is the kind of thing that would become a hit on YouTube.

I'm back on my feet and feeling determined to get some of the high cherries. It takes some doing — "Get 'er done" — but I manage to get the super-long ladder propped on some branches a few feet away from the trunk. I hit a real treasure trove of fruit. Lurking behind some leaves are clusters of 25 to 50 cherries. The basket is filling fast and I notice its weight around my neck.

Chuy's happy return

It's about 11 a.m. when Chuy pulls up in a van. There are six or seven lugs full of cherries and I'd guess I did at least two of them, maybe three. The other guys have more lugs full by another row of trees. Chuy samples a few of the cherries and declares them to be excellent. "You won't get this quality at the chain stores," Chuy says.

He is in a chatty mood, and tells me he figured I would have left a couple hours ago. Chuy, 35, has worked at Watt's for 20 years. He said he knows all 40,000 trees in the 250-acre Watt orchards, which are mostly apples. He knows the strong branches, the ones that will hold a ladder.

He says local Americans have come out to pick fruit before and rarely has anyone stayed more than an hour. The ladders and the hard work scare off locals, he says. Many of the Hispanic workers don't like the ladders either. But they get used to them, and learn how to angle the ladders to get all the fruit on the trees, he tells me.

Chuy agrees that cherries are challenging to pick, but he says apples are far more difficult. They are heavier for one thing, and they don't come off the tree very easy. The pickers have to be careful not to squeeze the apples or they'll bruise and won't sell. Every apple has to be carefully set in the basket, or else the fruit will get brown spots, he says.

The pickers can have 40 or 50 pounds of apples swinging from their necks and backs when they're on top of the ladders. Last year Chuy says the local apple farms couldn't get enough pickers. Watt's didn't pick any apples from an 18-acre orchard because of the labor shortage, Chuy says.

I tell Chuy that I've maimed many branches, either snapping them off while grabbing cherries or scraping them hard with the ladder. One of the thicker branches cracked during an aggressive ladder move.

"It's impossible to leave the tree unchanged," Chuy tries to assure me.

He and the other workers will trim the damaged branches in the fall and winter, trying to seal off the wounds.

I tell Chuy I am struggling to securely prop the ladder up on the outer edges of

Jesús "Chuy" Vallejo gives Tom Rivers cherry-picking tips. Vallejo says to be careful not to snap the branches.

the trees, where there's just the end of branches, without a strong support. He shows me how it's done. He takes the ladder and pushes it almost straight up one side of the tree. This requires forearms of steel. The many smaller branches seem to hold the ladder in place. He again makes a quick climb up, does a little shake, and pronounces everything safe and secure.

He heads off with the lugs of cherries, which will be sorted at Watt's and put in the cooler.

View from the top

The new ladder placement reveals a goldmine of cherries. It's still a little tricky and slow because I'm holding onto the ladder or a branch with one hand while picking with the other. My arms are getting tired. I decide to climb to the top of the ladder and survey the scene. There's got to be 40 or 50 acres of fruit trees. I see lots of leaves with tiny apples emerging. I can see the Watt country market and the train station a quarter-mile off near Route 98.

It's mostly quiet. There's some chatter and joking around with the Hispanic guys, who are about 50 to 75 feet away. One of them breaks into soulful singing every 20 minutes or so — Spanish ballads out of nowhere. I'm glad the guy is enjoying the job.

It's noticeably windier at the top of the trees. I've told many farmers in Albion and Gaines I didn't think we had enough wind for the giant turbines. I rarely feel much more than a breeze in the village, where I live. But the farmers tell me it's windy out in the country. I guess they're right. The two towns were being considered for a wind project with at least 55 turbines, but Gaines rejected the plan. There's always the chance the state could usurp the towns and site the turbine projects. So who knows what will happen.

I'm clinging to the ladder and taking a breather when Chuy shows up again, announcing it's lunch time. He takes the four other guys in a van.

Lunch confessional

I return to my trusty Tercel and head to Watt's market, hoping I can get a free ice cream cone and some details about these cherries.

Karen and Chris Watt are both there, and they offer me a salad. Karen notes Chris has lost more than 50 pounds with his healthier food choices. He hasn't lost his edge, though, and remains one of Albion's most intimidating people.

But I know Chris really isn't that scary. This guy wears a conductor's outfit on the weekends when he gives tours of the farm on the train, often to little kids.

I immediately complain about the ladders, telling them no one informed me I'd be climbing 25 to 30 feet. Chris corrects me: "They're 16 feet."

Well, it felt like 40 feet.

He tells me he went easy on me, by only having me carry a half-bushel basket. The full bushel ones weigh at least 40 pounds and surely would have killed me, Chris says. (I told you he's a softy.)

I tell Chris I may have done a little damage to the trees — that a few branches snapped off in the process of my work. Chris puts down his fork and seems a little agitated.

"Anytime you wound a tree you have the possibility of insects and disease getting into it," he tells me.

I decide then to not mention the rather large limb that cracked during my ladder battle. It's possible it was a pre-existing injury. Chuy already has been notified.

I ask Chris why he's in the cherry business because I don't hear much about cherries. The local fruit crop is dominated by apples.

An early crop

Watt Farms has 3 acres of sweet cherries, just a sliver of the overall orchard. The cherries are one of the first fruits to ripen. They're ready when the farm market season takes off in July, when Watt supplies four markets, as well as selling them from the farm's home base in Albion. Not many growers have cherries, a fickle fruit that doesn't stand up well to Western New York winters or the occasional frost that hits in May, Chris says.

Chris would like to grow more cherries but he says it's hard to find labor to pick them in late June and July. He pays the workers by the hour so they don't rush the work, maiming trees and picking fruit that's not quite ready or is substandard.

A worker from Watt's empties a 20-pound basket of cherries into a lug.

I ask him about the tall trees. Can't you use dwarf cherry trees, the

Tom Rivers picks cherries near the top of a tree in Albion, N.Y.

shorter ones like many of the apple varieties? They seem to peak at 10 feet in height. There are small cherry trees but Chris says they wouldn't be much of a match for Western New York winters and the many pests that prey on the trees.

"I don't know anyone who has a successful orchard with dwarf cherry trees," he says.

He tells me the farm is raising the price for a quart of cherries by $1 to $4.95. He expects some complaints from customers. The cost of sprays, fuel, labor, utilities, replacement parts, tractor tires, "anything we touch," have all gone up dramatically, by an overall 26 percent from last year, he says. The $1 more per quart just covers some of those increases. I suggest he charge $100 a quart.

Lunch is over and I tell Chris I want another shot at some of those high cherries. It's clearly my ego talking because I'm more than willing to call it a day. I tell him I'll be out there maybe another hour.

Climbing with confidence

I head back to my spot where Chuy placed the ladder so perfectly. There are cherries within an arm's reach up and down the ladder, which suddenly seems diminished in size, now that I know it's only 16 feet high. I try a new technique Chris suggested with both feet wedged in the ladder, with my back to it. That way both arms are free, and I can see what I'm doing. I'm truly enjoying myself with this new approach. I'm not hurting the branches and I'm becoming proficient in getting the cherries by their stems. Everything feels right in the orchard world as long as I don't look down.

After getting everything I can reach, which takes about an hour, I decide I'm not quite ready to go home. I want to set the ladder myself and get some more high cherries. I lower the ladder, move it over a few feet and slide it up through some branches.

There are cherry clusters all over, and I delight in working up the ladder. The sun is out in force after the overcast morning. It feels like you could cook eggs on the back of my neck. I scoot down to where some leaves offer shade.

I notice some bugs walking along the branches. I start to feel creeped out by the leaves that keep brushing up my pant leg, and on my neck, and on my arms. It feels like a centipede is crawling on me. I decide to work the ground floor again.

I fill up another basket and nearly have two lugs full from the afternoon. I just need a little more to fill a second lug. So I make one more stab on the ladder, moving it one more time for a bonanza of cherries. I'm at it for another half hour, filling the basket before I call it a day. It's almost 4 p.m. I've picked five or six lugs, nearly 200 pounds of cherries.

I feel like a conquering warrior. But next to the cherries lurks rows and rows of apples, a greater challenge for a novice picker. I will meet up with them in a couple months, and this time I will have some mastery on the ladder. It's a rendezvous I relish.

Chapter 5

Picking berries is picky, prickly, hot, sweaty work

Maestro, left, empties a bucket of blueberries into pint containers. Maestro is a legend at the farm for his incredible output. (Photos by Nick Serrata)

POINT BREEZE — The guy's name is Lino, he's tall, about 20, and appears to have eight arms.

On a Saturday in mid-July he seems to be the star picker in the blueberry and raspberry fields at Brown's Berry Patch. Lino is often next to me along the rows of berry bushes. Then he isn't.

While I methodically work a bush, Lino weaves around me. In my peripheral vision, he looks like an octopus, grabbing berries in front of him, behind him and to the sides.

I am particularly cautious with the raspberries. Those bushes have prickers. I gingerly reach my arms in, trying to grab a berry or two. Lino just walks into the bush, the prickers poking his midsection and arms. He seems unfazed.

Lino is one of 13 workers at Brown's through the federal H-2A program that allows legal temporary farmworkers to be brought into the U.S. Brown's currently has nine workers from Mexico and four from

Jamaica. When the 200-acre fruit farm gets busy with the apple harvest in September, it will have 30 workers through H-2A. The farm pays their transportation to Point Breeze and then back home, and also must provide free housing on top of the pay.

The workers are guaranteed at least $9.70 an hour. But if they're fast pickers, they can make more money. Some earned $18 an hour picking cherries in recent weeks, farm co-owner and orchard manager Eric Brown tells me.

Lino doesn't appear content with the base pay. He frequently fills a plastic pail of berries, then walks to the end of the rows to fill pints with either blueberries or raspberries.

The pints fit into a cardboard flat holding 12 containers of berries. A good picker can fill 20 pints with blueberries in an hour, José Rodriguez, the crew leader and a 10-year employee at Brown's, tells me. He keeps track of each worker's output. The flats are numbered from 1 to 13, showing each worker's production.

José also loads the flats of berries on a Gator tractor and runs them to a cooler at Brown's. Eric says the berries should be cooled within two hours of being picked. That way they're fresh when they're delivered to Wegmans or Tops the next morning. At the Brown's farm market on Route 18, picked blueberries sell for $2.99 a pint while raspberries are priced at $4.29.

Berry qualities

The raspberries are more time-consuming to pick, despite being about twice the size of the blueberries. The raspberries have to be gently slid off the branches. If they're not ripe, they don't budge and separate from their white core. And don't squeeze too hard or else the berry will turn to mush, especially on a hot summer day.

A good raspberry picker can fill a dozen pints in an hour. I think I did three.

The blueberries are easier. You just roll them off the branches. Each blueberry bush seems to have about 1,000 berries — and no prickers. The berries are at a range of readiness. There are emerging green berries that are like hard little stones.

Others have a purple color. They should be ready in a couple days. Others are blue, but the bottoms are still purple. Give them another day. If you pick them, they'll have a sour taste. "They'll turn your tongue backwards," Eric says.

You want to pick the dark blueberries. Be patient and wait on the ones almost ready. If you're a picker at Brown's, you'll be passing back through the berry bushes in the next day or two. Get today's almost-ripe berries then.

Eric tells me the farm used to have many high school kids pick berries. But they weren't very selective with their picking, grabbing too many berries that weren't quite ripe. Too many sour berries and the grocery stores will reject the full load of berries. "With high school kids you're paddling upstream," Eric says.

U-pickers

The farm's picking pros work berry bushes within earshot of Brown's farm market, petting zoo, playground and a new giant jumping pillow, among the many attractions to the retail outfit Brown's has built up in the past 25 years. The farm also has a busy U-pick operation where regular folks can pick their own cherries, berries and apples.

Many of the U-pickers cross paths with the picking pros. I hear one grown woman exclaim at the sight of a blueberry bush. She has never seen one before. One young father is picking blueberries with his daughter, who seems 3 or 4. She has picked many green berries. "Honey, the green ones are yucky and the blue ones are yummy," I hear him tell her.

The U-pickers seem to come and go, maybe spending a half hour out with the "professionals." The amateurs pull wagons loaded with cartons of berries. The pros have plastic or metal pails hanging from their shoulders.

There is one particularly rowdy U-pick group. About 30 India natives, most who now live near Niagara Falls, have been coming to Brown's for years. They are delighted to be in the orchard. Some of the young Indian kids pull their fathers on wagons to the cheers of the onlookers. These people can't stop laughing.

One of the fellows from India comes over near me and picks several

Tom Rivers, center, joins a crew of Mexican workers in picking raspberries at Brown's Berry Patch. The farm, which started in 1804, is the oldest in Orleans County.

blueberries. They aren't quite ready. They have purple bottoms. The guy eats them all, and he knows something isn't quite right. I tell him to pick the darker blue berries. He does and he seems satisfied.

I notice one of the U-pickers pulling a wagon that has about a dozen pints of blueberries. Oddly the guy also is hauling three or four branches from a bush.

Pickers brave the prickers in the raspberry bushes. The group includes, from left: José Rodriguez, the crew leader; Tom Rivers, reporter; and Lino, one of the farm's most prolific workers.

Muggy day

I just go about my business on an awfully hot and humid day. By 11 a.m., I can barely stand the smell of myself, an unpleasant brew of sweat and sunscreen. José, the crew leader, stops by and he tells me he doesn't like these hot days. But he says berry picking isn't too bad.

The crew has been alternating among strawberries, cherries, blueberries and raspberries in recent weeks. The strawberries finally are done. José says they wear out the workers because of all the bending to pick them. But the blueberries and raspberries are on bushes about 4 to 5 feet high. You do have to squat down to get some of the lower ones, but you can grab many of them without bending. I am relieved no ladders are involved. To get cherries about two weeks ago, I had to climb a 16-foot ladder.

José says the ladders will come out for the apples. He says apples are a difficult chore because the workers are carrying 40-pound bins and they're going up and down the ladders constantly.

Worker worries

José is concerned about getting the apple crop in. He knows 10 of his friends who worked at local farms won't be back in the fall. They instead are working in New Jersey where José said there is less immigration enforcement.

José Rodriguez gives Tom Rivers some pointers on picking berries. Make sure blueberries are ripe, José advises.

José, 40, has worked in the United States for 20 years. He started out in Dallas, Texas, working in a restaurant for five years. The country boy from Mexico didn't like the city life and switched to farm work, first in Arkansas.

He came to Western New York because a cousin was working at local farms. José worked at other farms before settling with Brown's. He thinks cutting cabbage and picking cucumbers are among the hardest farm jobs because of the constant bending.

He likes the Brown fruit farm — no cabbage or cucumbers. The owners provide the workers with laundry machines at the houses and a labor camp. The farm also provides the workers rides into town for groceries, money transfers and other needs, José says.

The workers, although they're all legal through the H-2A program, don't like going into town, José says. Many have stopped attending church because of the stepped-up immigration enforcement in the past two or three years.

José has a legal Green Card. About a week ago he was returning from a two-week visit with his family in Mexico. He was stopped by immigration officials in Atlanta and questioned for two hours. He missed his flight and had to wait another day to fly back here. He wonders if farms will get enough labor to bring the crops in.

Eric Brown believes his farm has enough workers now and in the fall because of the H-2A program. "Without the H-2A workers we couldn't harvest our crop," he says. "You can only get so mechanical."

A blueberry harvester can't discern the color of the berries. The color shows which ones are ready to be picked, Eric says.

Hail curse?

I had intended to go to Brown's the previous Friday to pick the berries, but it rained in the morning. The farm doesn't pick wet raspberries because they are then more prone to mold. So we put it off a day.

So here I am, a day later, and rain is threatening. About an hour after working on the raspberries, the sky gets dark. A few minutes later, we are bombarded with heavy raindrops. The pickers all book for a loading truck in the orchard that allows all 10 of us to squeeze in.

It is raining so hard I fear hail. I had already told Eric and his brother Bob Brown that some people in the local agricultural community have complained, half-jokingly, that I was bringing a hail curse on farms I've worked this spring and summer. (Neither of the Brown brothers found this prospect humorous.) About a month after helping plant onions at Triple G Farms on the muck, the onions were hammered with two hailstorms on June 16. The section I worked seemed to be hardest hit, farm co-owner Guy Smith has told me.

Watt Farms in Albion dodged the June 16 hailstorms. But two days after I picked cherries June 27, another hailstorm roughed up Watt's apples.

Rain respite

The rain quickly tapers off at Brown's and slows to a drizzle after about 10 minutes. The Mexican workers declare it lunchtime, and in the five minutes it takes me to walk to the farm market, the rain stops.

I am drenched. I find a spot at an outdoor table all by myself. I don't care that the seat is wet. I really don't want to be inside with all the noise and chaos from kids and out-of-town tourist-types.

In a few minutes I am joined by Eric Brown. I tell him I am treated well by José and the guys. I say I am thankful for the rain, because the humidity was starting to suck the life out of me. Eric admits it is a tough day to be picking. The previous week had been breezy, not hot or humid.

After eating a Julienne salad — I'm on a health kick — I am back out in the field. The raspberries are off limits for the rest of the day, but blueberries, because of their outer skin, are still in play. The sun is in its full glory, just sizzling. I apply my sunscreen for the third time.

Lino is still going strong. A few times he gets close to me, encroaching on my territory. He just steps in, makes a few grabs and keeps moving down the row. I decide to sit cross-legged under one blueberry bush and hit all the low ripe berries. It takes 20 minutes or more just to get those.

Workers head for an old delivery truck in the orchard during a hard rain storm. The truck is usually used as a loading area for the berries. Sometimes it's shelter from a storm. (Photo by Tom Rivers)

I welcome the shield from the sun.

As the sun bakes the orchard, the bees, flies and other loud-winged bugs seem to come out. I'm allergic to bee stings, so I feel a little antsy. One bee sounds like a World War II bomber plane when it flies at my ear. I let out some kind of noise and do an awkward little dance. This draws amused looks from the Mexican guys.

We plug along for the next two hours. One guy, a tenor, sings a few songs. Another guy displays his gift for belching the musical scale.

Meet Maestro

Bobby Brown, Bob's son, stops by around 3 p.m. to grab the flats of blueberries and to replenish the water supply. I marvel to him about Lino. But Bobby, who's two years out of Cornell, says there is a more prolific picker, a guy they call Maestro. He picked 31 pints of berries the first hour after lunch. I picked 14 or 15 pints in 2 1/2 hours.

Maestro is barely over 5 feet tall. He's a few rows away from me and I can't get a good sense of his picking style. But he is clearly a legend at the farm.

I tell Bobby I'd like to get another six pints. I already have two or three picked in my pail. When I get six, I'm done.

Fading fast

About 15 minutes goes by and it hits me. I'm dog-tired. The six-plus hours have sapped my energy. I dump my berries in the flat, but I only have four pints. I should get two more.

I take four enormous gulps of water and go back to picking. I look for mother lodes of ripe berries, but the bushes have been fairly picked over. I spend 20 minutes more, hoping I've got the two pints. But I've got barely more than one.

I will myself back to the bushes. Ten minutes later I think I'm done, but I'm still a little short. I do one more cross-legged sitting session. I get 150 or 200 berries, and it's enough to top off the six pints.

It's a slow walk back to the farm market. I thank Bob Brown for letting me stay. There's a big crowd in the market, and I see Medina Mayor Adam Tabelski. We chat but I'm losing my ability to be social. I see more people I know and give them a quick "Howdy."

I just want to get to my car and go home. I've almost reached my Tercel when I hear a lady yell out "Tom!" I don't turn around. I get in and drive. Whoever that was, my apologies. I didn't have enough left in the tank.

Chapter 6

Showing animals calls for attention to details

Kaitlyn Dresser scrubs a Holstein's belly as part of the early morning wash for the animals. (Photos by Tom Rivers)

BATAVIA — Kaitlyn Dresser is doing a serious deep knee bend, her hands squeezing soap suds into the legs of a Holstein. Then she is bent way over, her right arm darting back and forth over the heifer's belly, giving it a scrub.

Kaitlyn, 19, spends 10 minutes washing this animal, getting any dirt or manure off its hair and skin. It's just before sunrise on a Thursday in mid-July. The soap will give the recently shaved heifer a nice shine for when the public walks through the barns at the Genesee County Fairgrounds in a few hours.

The Holstein will be on display all week at the fair. But her main event will be on Saturday, the fair's conclusion, when 126 animals, from 300-pound calves to 1,300-pound moms, will be judged at the three-county Nioga-Genesee All Breeds Show.

There will be lots of pampering for the animals, and lots of work for the adults and teen-agers who will prep the calves, heifers, yearlings and cows for the show that runs from 9 a.m. to about 2 p.m.

"They get treated better than we do," Kaitlyn says, smiling while washing one of the Holsteins. The sun isn't up yet, but Kaitlyn and the team caring for the Holsteins rose before 5 a.m.

After she gets the cow soaped up, Barry Flansburg hoses off the suds and finishes the wash. Then Kaitlyn leads the animal back to the barn. I help a little bit, washing the Holsteins and leading them to the barn. But I don't do too much. When I arrived Thursday at about 5:15 a.m., the cows had already been milked.

Flansburg, the Nioga show chairman, agreed to let me show one of his Holsteins at the open-class show, when adults can get in the ring. I'm basically clueless about this despite watching and reporting about livestock shows for a decade. I've never tried it, and I admit to anxiety about the whole thing.

I've already had my feet stepped on hard by two cows, been yanked and pushed around by the animals, even the small ones, and that's just in the 25-yard walk back from the wash rack.

Little kids do it

But I assume it can't be too bad if little kids are showing these half-ton animals, including the bucking steers. I should be tough enough to do this. But I'm still nervous. Wednesday morning I saw a 10-year-old boy get led to the wash rack by a heifer. He was hanging on to the halter while the animal dragged him about 20 yards.

All of the kids seem to walk with the animals as if bracing for a grenade to go off, waiting for the Holsteins or steers to go crazy. I'm glad I still have some spring left in my knees because some of these heifers and cows will make a quick break to the right or left, or they'll start running when they see their pile of hay and other food.

It gets easier as the week goes on, the veteran dairy showmen tell me. Some of the animals need a day or two to adjust to the surroundings, the new setup for milkings and washings, and their noisy neighbors at the fair — the goats, horses and chickens.

The champ

Some Holsteins just don't get it. I watch Elton Shuknecht of Elba lead a cow to the wash rack. The animal takes two or three steps, then stops and looks around. The goats and horses are up and making a racket. The cow takes in all the sights and sounds, and goes at an awfully slow pace.

"She's not used to it," says Shuknecht, president of the Nioga Holstein Association. "I hope she gets used to it real soon."

Barry Flansburg trims the hair on a calf, one of Flansburg's many duties during the week. He is chairman of the Nioga-Genesee All Breeds Show, which includes 126 animals.

The other Holstein aficionados say Shuknecht is well known for bringing at least one animal to the fair that has never been on a halter, never seen a show ring and perhaps never left a dairy barn. Shuknecht can usually get these fair "rookies" into show shape by the end of the week.

But some animals just won't take to the halter. They'll stand still in the barn, like enormous anchors, refusing to move. Shuknecht may decide they're not show material.

Shuknecht, 52, has had the grand champion Holstein for the past three years at the Nioga show. He works as the dairy manager for Torrey Farms in Elba. He raises registered Holsteins on the side. It's his hobby. He enjoys the challenge of breeding better animals, those that produce more milk and have better bone and udder formations.

"I tell my wife we could have an RV, a cottage or a boat," he says. "But this is a passion more than anything else. This is the thing we do during the summer."

Shuknecht has a lineup of 15 registered Holsteins on one side of a fair barn. He is thrilled to return to the barn late Wednesday morning and see all the Holsteins lying down. They look ready for a nap. Shuknecht says the goal is to keep the animals comfortable and content during the fair. It's an around-the-clock job.

"I think it's a lot of fun," he says. "Maybe people think I'm strange for saying that."

A crash course in showing

Flansburg, 51, has three kids, ages 18 to 23. They showed Holsteins and other animals once they turned 9 and could participate at the Orleans County 4-H Fair. Flansburg's son Ben, 23, won the fair's grand master showman title four straight years. His sister Amanda, 18, won it last year. Chris, Amanda's twin brother, won the showmanship award at the Erie County Fair last year. If I'm going to learn, these are good people to hang out with.

Chris gives me a crash course in showing Tuesday night, behind the barn as the sun is setting. When showing Holsteins, grab the halter with your left hand, hold its head up, and walk back slowly. Keep your eyes on the judge. When the judge is looking at the animal, probably from its left or right side, try to position the cow with one back leg stretched back. That way the udder can be showcased, Chris advises.

If it's a calf, heifer or yearling — Holsteins that haven't had babies yet — the showman should try to emphasize the back and other composition of the animal. Try to get the judge focused above the yet-to-fully form udder. When Chris and the other showmen trim the heifer's hair, they leave a top line about a half-inch high along the back. With cows, the showmen will trim the hair on the udder very close to the skin, trying to show off the udder's veins.

Tuesday night I work with a Holstein that has been through the show experience many times. It easily follows my lead. Chris tells me to just back up slowly and pay attention to the judge. He pretends to be the judge, squinting sternly at various animal parts.

Chris Flansburg, right, gives Tom Rivers some advice about showing animals. The main thing: Stay calm. (Photo by Nick Serrata)

Chris says to try to get the animal's front legs parallel. If one is trailing, just give the Holstein a little tug with the halter. She should take a small step with the other one.

You also may want the Holstein's ears to stand up when the judge is near. Chris says to just whisper to the animal and her ears will perk up. The big thing, Chris stresses, is not to act panicky or angry. The Holsteins sense that. Just stay calm, he says.

Barry Flansburg hoses off the suds on a freshly washed heifer.

And another thing: Wear sturdy shoes. The heavy animals may step on your foot, put all their weight on it, and then turn in the process, almost grinding a human foot. Chris suggests some thick boots.

Whites

There isn't too much pressure on me for the show because I'm not being judged. It's all about the animal, Barry Flansburg tells me. Just lead the animal in the ring and don't do anything crazy. Let the focus be on the animal.

He says it shouldn't be too hard for a novice showman unless a heifer is in heat. "Then all bets are off," he says. That animal will be difficult to lead and keep under control.

I'll need white pants and a white shirt for the show. Barry loans me a pair of his pants. They're pretty long for me. "Just roll 'em up. That's what I do," he says.

Chris says he likes the white outfits. "It's classy," he says. "It's not jeans and a T-shirt like everything else. It makes you look a little neater."

His father spends most mornings at the fair hosing the Holsteins after his daughter Amanda and Kaitlyn Dresser lead them to the wash rack and get them soaped up. Barry Flansburg also needs to manage the show, preparing programs and checking with the 50 different exhibitors who will have animals in the Nioga show. He may stop by the town halls in Barre, Oakfield and Byron, where he is town assessor.

He finds the fair a great way to stay informed on land values, especially in the agricultural sector. He sees many farmers at the fair and he gets an earful about the ag economics in the community. "This is so much better than sitting in the office," he says. "People feel more free to talk to me here."

Flansburg is covered in Holstein hair while talking to me Tuesday morning. He has the job of trimming the animals, shaving their hair for the show. He has several clippers of varying blade heights for the job. He lets me try and the clipper quickly takes off gobs of hair on the calf's side. It's easy to do on the sides, but much harder working around the bones, Barry says. There's all kind of nooks and crannies behind the tail, between the folds of skin, where the brisket joins the shoulder. He says it will take at least an hour to trim the animals, but probably longer because he likes to chat with just about everyone who walks by.

Team approach

The Flansburg family started raising registered Holsteins 10 years ago when Ben won a free registered Holstein calf, part of a state-wide youth competition. That Holstein then led to another. Now the family has 12 and spends four weeks in the summer traveling to Holstein shows.

They each have found their own niches in working with the animals during the week. Barry gets the animals trimmed and hosed off. Amanda moves them to the wash rack and gets them clean with soap. Chris gets the cows milked early in the morning and then around 5 p.m. He also manages "the pack," the layer of straw and bedding for the animals.

Karen Eick spends the morning cleaning cows for her friends at the Genesee County Fairgrounds.

Everyone watches the animals during the day, making sure they are fed. They also quickly clear any pee or poop off the straw. Flansburg says he wants the public to see clean animals.

The family also recruits some long-time 4-H friends to help during the fair week. Kaitlyn, a college student who wants to be a nurse, has her own Holstein for the show. She is helping care for the Flansburg animals as well.

Next to the row of Flansburg animals, long-time friend and fellow Barre resident Andy Beach, 23, has 11 Holsteins in his care. He asked family friend Karen Eick, 25, of Medina to help with the animals this week.

Eick, a substitute teacher and coach at Medina, welcomed the chance. She doesn't mind the very early mornings and the late nights.

"I like doing this," she says Wednesday at about 7 in the morning. She is almost done cleaning one of the cows. "It brings me back to when I was young at the fair." Flansburg says he appreciates Eick for showing up and working hard.

"You know you have a true friend when you can call Karen up and say, 'You want to come and wash cows?' and she says, 'Yes!'"

Chapter 7

Succeeding in the show ring

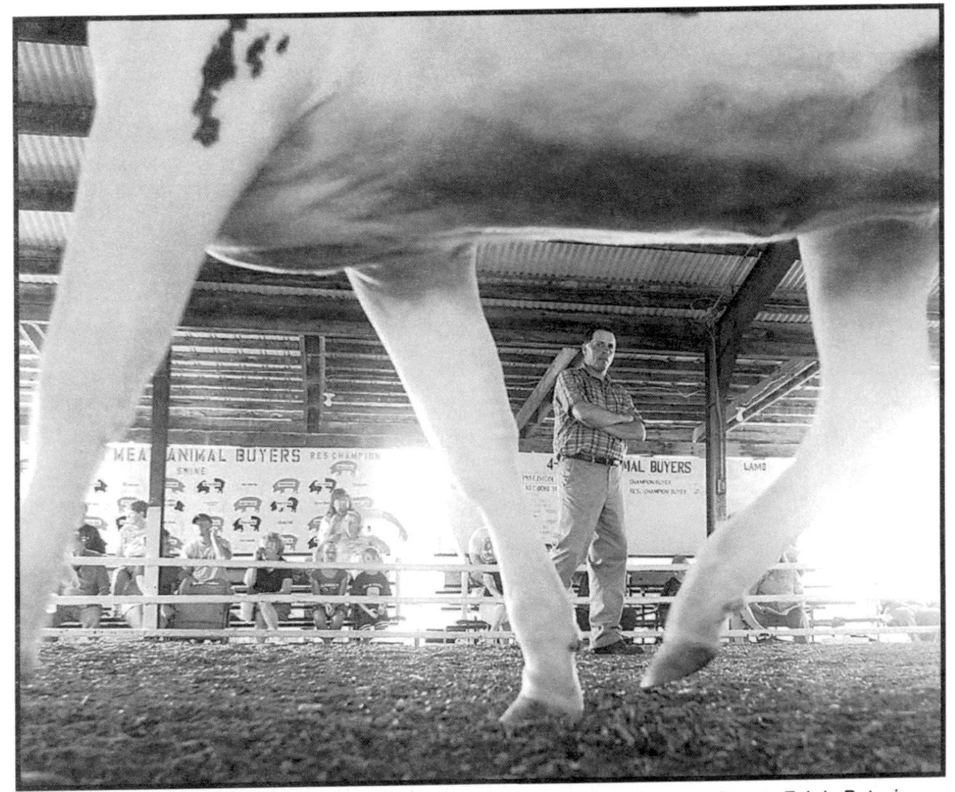

Judge Adam Liddle watches a Holstein move in the show ring at the Genesee County Fair in Batavia, N.Y. (Photo by Mark Gutman)

BATAVIA — Just moments before my debut in the show ring with a registered Holstein, the 800-pound heifer next to me sneezes, unleashing yucky goo all over my right arm and on the fine animal I am holding by a halter.

All of my preparations during week — the practice walks with the heifer, trying to get her to hold her head up, keep her back straight, act calm — did not touch on this predicament. I am happy the heifer doesn't seem to care. So I guess I shouldn't either. I just wipe the slop on the way-too-long white pants I borrowed for the day.

Then the judge tells me to start leading the pack of five animals in the Holstein Summer Yearling Heifer class. I really don't want to be the first in line. I hoped to just follow the leaders, the experienced showmen in the ring. But judge Adam Liddle tells me I'm first. So I have to get out there with the heifer.

Before we start to walk, I hear a 4-H'er yell, "Good luck, Mr. Rivers."

I also hear a member of the Swine Club heckle me about showing a small heifer, rather than an enormous milking Holstein.

I tighten the halter around the heifer, and start to backpedal, leading her while walking backwards.

Remember this

There are two tidbits I seared into my memory before the dairy show, answers to questions the judge may throw at me. Chris Flansburg, 18, of Barre has quizzed me a couple of times already this morning.

"Father?" he asks.

"Dundee," I answer.

"Date of birth?" he asks.

"Aug. 14, sir," I say.

Chris has agreed to let me show his heifer, part of a 12-animal registered herd he owns with his twin sister Amanda. The Flansburg family of Barre has been working with registered Holsteins the past eight years. It started when Chris's brother Ben, now 23, won a Holstein calf as part of a youth development program through the New York Holstein Association.

The Flansburgs, including their father Barry, are slowly building their herd and working on the animals' bloodlines. Chris says it will take years to develop superior Holsteins. His family is new in the business, and they readily admit they aren't quite in the same league with established Holstein stars such as the Reynolds family in Darien, the Welkers in Lyndonville and Elton Shuknecht of Elba, just to name a few.

Chris has given me showing tips throughout the week, and I've also watched all the work the showmen and their supporting cast put into getting the animals ready for the show.

The cows are milked at about 5 a.m. and then again at about 5 p.m. They are washed daily with soap and hosed down. They all get shaved, with the hair-cut artists leaving top lines of hair along the back bones and fluff-balls of hair at the end of the tails.

The animals are steadily fed. Their poop quickly gets snatched with a pitchfork and deposited out of the barn. That way the Holsteins won't smear any poop on themselves when they lay down for a rest. The showmen also are mindful of all the public relations going on during fair week, when youngsters and adults stream through the barns, looking over the animals.

Leading the pack

The heifer and I are going at the desired slow pace, leading the pack of five summer yearling heifers. I backpedal close to one of the rails on the show ring, watching the judge. He zeroes in on the heifer, analyzing her legs and her shoulders while she moves. He comes up to me.

"What day was she born?" he asks.

With all the confidence I can muster, "Aug. 14."

He taps her on the back and moves on to the next one.

Tom Rivers tries to follow the advice of the experienced showmen by staying calm. The animal can sense any jitters from the handler.

I complete a full circle in the ring and I'm not sure what my next move should be. Oakfield dairy farmer Jonathan Lamb is along the edge of the ring and he kindly suggests I make another circle.

I keep moving and get about halfway around when the judge stops the action. He points to a heifer and says "first." He comes up to me and says "second." He moves down the line.

I watch the first heifer start to line up and I follow her, except I'm on the wrong side, the "first side," when I should be on the "second side." The judge comes over and points to the other side. He doesn't seem too annoyed with this little delay. I get the heifer on the second spot and there we all are, the five heifers lined up with their backsides facing the judge.

He studies each one. He comes up to me and tells me he likes the rump on the heifer and the wide spacing between her ribs. But her shoulders aren't very wide and they're on the small side. He says some other things I really don't understand.

Adam Shuknecht, 16, of Elba is standing next me with his heifer, listening to the judge's comments. "I bet this sounds like gibberish to you," he says. He knows I'm a reporter, just trying my hand at cattle showing.

I tell him I'm clueless about what the judge said and what it all meant. "The main thing is you don't want them to slobber all over you," Adam says. His heifer is the one that sneezed just before we entered the ring.

Kaitlyn Dresser, 18, tries to set up a red-and-white Holstein she is showing at the Genesee County Fair.

Decent showing

The judge makes his decision, giving first place to the heifer owned by Peggy Bennett of Barre. The heifer I showed comes in fourth overall. Chris seems happy. It's not last place, and he is going against many well-established farms.

His heifer was the youngest in the ring, with one of them 10 weeks older. The older animals had more time to grow and develop "dairy characteristics."

I catch up with the judge, Adam Liddle, during the lunch break and he explains his critique of the heifer. The big rump will help the heifer when she is giving birth. The wide-spacing between the ribs often is a good indicator the heifer will be a prolific milk producer, Liddle tells me.

I ask about the narrow and under-sized shoulders. Liddle says that's really not a big deal. It won't hurt milk production or make birth harder for the animal. The wider shoulders just look better.

"It's a beauty pageant thing," he says.

He says the heifer is a high-quality animal.Chris confirms the opinion. He thinks the heifer could be special. Her mother is rated "excellent" by the American Holstein Association, the first Holstein the Flansburgs have with such a rating. Her children and her children's children will carry those bloodlines and boost the quality of the Flansburg herd.

Hectic

There are 126 Holsteins, from small calves to milking mothers in the Nioga-Genesee All Breeds Show. The action starts at 9 a.m. There are calves, heifers and milking cows coming and going constantly from the barns and big tents. It's hectic, and it gets more chaotic when it's time to judge the groups — dam and daughter, dairy herd, breeder's herd, best three females, and others.

Sometimes a farm will need five animals in the ring at once, but the farm might only have two or three people who could go in the show ring. That leaves the farm in need of at least two bodies. They also should have someone in the barn, getting the next line of animals ready.

Doug Welker of Lyndonville comes up to me and says my services are needed. He wants me to join his two teenage sons, Riley and Zack, in the ring, along with the boys' mother, Paula. We'll all be showing dairy cows.

I eagerly agree, and I am thrilled to get a red and white cow. There's not too many of those at the fair. I go to the Welkers' spot in a large tent and Doug hands me the halter that is secure to the cow. This girl is used to the show scene. She heads right for the ring.

She never gives me any trouble and I don't have to think too much. I just follow Zack and his cow.

The judge marvels at the quality of all the dairy animals. He has a tough choice. He praises the Welkers' cows and gives them third.

We leave the ring in a pack of about 20 people and 20 enormous cows. There is a new set of animals already crowding in. I get squeezed between two cows and feel a much-needed chiropractic adjustment.

I am ready for more action. Chris has entered the dam and daughter group, showing a heifer or a cow and her daughter. I get a return engagement with the summer yearling heifer, while Chris leads the mother. I just follow him and try to line up the heifer next to the mother when the judge studies their features. Liddle moves quickly through the pack because he has already seen these animals earlier in the day.

There are 11 pairs in this group and Chris places eighth. "It's not last place," he says.

He thinks eighth is a good showing. Some of the dam and daughter pairs are fully grown cows, their attributes as milking cows plain to see. Chris said the judge can't score a heifer or a calf too high when they're up against milking cows.

Tom Rivers backpedals with a heifer during his debut in the show ring. Judge Adam Liddle inspects the animal's "dairy characteristics."

I get one more shot with Chris when he enters the

breeder's group, a collection of five animals bred at a farm. This is the first time Chris has entered the breeders. You need at least two milking cows to be in the group.

Chris has cows, heifers and a calf in this group. I get the calf and she ends up being the trickiest to lead. She's not yet used to show business.

She steps on my feet four or five times when I'm backpedaling. Because she is smaller, with a shorter neck, I am closer to her body and feet than with the bigger animals. I opt against the backpedal approach and instead lead her from the side. This solves the problem. She is pretty easy to direct around the ring and feels like a little dog compared to the bigger Holsteins. I just follow Chris and join him in the lineup of five animals.

He finishes fourth, the lowest of the breeders, but again Chris doesn't take it as bad news. He is competing against long-established operations. He knows his herd will only get better. He just needs time.

"When you're up against people who been breeding for years and we've just started eight years ago, you can't expect to win them all," Chris says.

Chris will start his college studies this fall at Iowa State University, pursuing either a degree in power machinery or systems engineering. He will design tractors and other farm equipment, or design barns, grain bins and other buildings for agriculture.

He expects his career will help him in the breeding business, giving him added cash to help develop better dairy animals. He thinks the Flansburgs are making progress with their Holsteins.

"Just being here competing against these other farms is a lot more than most people could do," he says. "This is a lot more than just grabbing animals out of the free-stall barn."

Chapter 8

Mixing it up with organic veggie variety

Tom Rivers joins a worker named Guilleminia in picking peppers in early August at Porter Farms, an organic fruit and vegetable farm in Elba, N.Y. (Photos by Rocco Laurienzo)

ELBA — It's 7 in the morning and my new co-worker, Guillermo, sits on a concrete slab outside the main barn at Porter Farms. He's sharpening a knife.

One glance at the razor-sharp, 6-inch blade tells me I'm in for it: an intense day in the fields. Guillermo and I and three other blade bearers on an early August day harvest zucchini, two types of squash, sweet peppers, two kinds of lettuce and many green beans. We need our knives for most of the work, doubled over, hacking away at the stems and stalks.

Our first stop: zucchini. We are like hunters, slicing stems from rather large zucchini plants with enormous open orange buds that attract swarms of bees. Porter Farms wants the zucchini cut at about 6 to 8 inches long. They're more tender that way than the long, fat zucchini I remember my mom using to make bread when I was a kid.

There were five or six rows of zucchini, maybe 50 yards long each.

Totes, each weighing about 75 pounds, are filled with tender zucchini and squash. They are picked before they get too big. The farm leaves any foot-long ones to rot in the fields.

We check each plant, lifting the huge leaves in pursuit of the just-right fruit. (I always thought zucchini were vegetables, but I was told by Katie Porter, a recent Cornell grad, that zucchini are, in fact, fruit. Go figure.) Each plant seems to have two or three zucchini that are ready. More will come in the next month or so.

The workers check the plants every two to three days. Zucchini can grow like crazy, especially on warm nights, and letting the fruit linger even a couple of days means they will be too big. We find a few foot-long chubby zucchini, ones that were missed a few days before. We chop their stems and leave them between the rows to rot.

Bees and buckets

We each have responsibility for at least one row as we move across the field. I admit an odd thrill using the knife, as if I had graduated from the minor leagues of picking berries a few weeks ago to working with the knife-wielding pros. This zucchini job is far more difficult than the berry picking, where you can stand upright and just give the fruit a little tug from its branches.

With zucchini, we are all stooped, searching for the perfect produce. I am leery of all the bees, but they don't seem to notice me. They are doing some serious pollinating with those enormous buds. I just go about the work quickly, lifting leaves and slicing off stems. The bees prevent any dilly-dallying on my part.

We zucchini hunters carry buckets in one hand and knives in the other as we collect the produce. It takes maybe 10 or 15 minutes to fill the bucket with about 50 zucchini. Then we unload them in yellow totes on the back of a pickup truck.

An American guy named Billy organizes all of the zucchini, maximizing the space in the yellow totes. Billy is in his 50s, hails from West Virginia and has worked at many local farms, including the past four years for Porter. He is a nonstop talker, who likes to pick on everyone and flirt with the ladies.

We have enough zucchini after about 45 minutes and then it's on to the sunburst squash, a fruit Porter added last year. Mike Porter, the farm owner, says he tries to add new vegetables, I mean fruits, each year because, "People want variety."

The sunburst squash tastes like yellow summer squash but it has a much more interesting look. They have flat and circular heads. They will make their season's debut with Porter's customers the following day.

Organic

The farm sells most of its fruits and vegetables to wholesale markets, including Whole Foods in Boston and New York City. But since 1996, Porter also has run an increasingly popular Community Supported Agriculture program, where people pay a subscription and get a 12-pound bag of produce each Saturday. The farm now has 575 CSA members from Buffalo to Rochester, including about 100 in Genesee County. They pay about $15 a week for a bag.

Mike says the CSA accounts for more than a third of the farm's produce business and he would like to see it continue to grow, perhaps hitting 700 members by next year. The farm had 100 CSA members in 1996 and gradually grew that business by 10 to 15 percent until the past two years when it went from 330 in 2006 to 501 the end of last year. It's approaching 600 now.

Tom Rivers searches for peppers to pick at Porter Farms.

"It's just had huge growth," Mike says. "People are pushing eat-local. There's been a lot in the news about it."

The more CSA customers, the less wholesale buyers Porter will have to deal with. Mike readily says he doesn't like working with many produce buyers, "who pit one grower against another."

The farm's fruits and vegetables are certified organic, meaning Porter doesn't use any conventional chemical sprays to fight insects and disease. Porter employs other tactics to ward off pests and disease, including using soaps and a water-based solution with fish.

Most of its plants are seeded in soil and then covered in long tarps of plastic, with holes poked through for the crop.

That plastic helps keep some of the pests in check. Porter also rotates crops and unleashes natural predators, various insects that will gobble up crop menaces. Porter often will set loose wasp larva and lady bugs to eat other insects that would harm the crops.

Because the farm doesn't use many conventional weed controls, the workers are weeding far more often than Porter's neighboring farms. Some of the weeds don't get yanked and they are prominent in the fields. Katie Porter, Mike's 22-year-old niece, warns against wearing shorts in the fields, because the weeds and big leafy plants will scratch your legs.

From field to sink

The sunburst squash is just starting to ripen, and it's not easy picking. Each plant may have one squash that's ready. The buckets aren't filling nearly as fast as with the zucchini. The squash has a tough stem and we still need to use our knives.

We finish checking the rows, and it's on to yellow squash. These fruit or vegetables — I'm not sure which — easily snap off from their stems. You can do it with a bare hand. I'm a little disappointed to have to put the knife away. It takes about two hours and we have our haul of zucchini and the two types of squash.

We drive 300 yards or so back to the main farm, where all of the fresh-picked produce will be washed in sinks. There are about eight of us unloading produce, washing it in sinks and then setting it in waxed produce boxes. Some of the workers are counting how much goes in each box. I'm basically oblivious to all the numbers being added.

I notice Katie, the recent Cornell grad in biological studies, preparing to hoist a tote of yellow squash. It looks heavy and I offer to help.

"I'm fine," Katie says, and she seems to easily carry the container of over-flowing squash to a table.

I figure I'll get the next one and I grab a tote. It must weigh 75 pounds, and I

Katie Porter, left, gets a box ready to be filled with zucchini while other workers rinse the produce that was picked only an hour before in a nearby field.

adjust my grip about five times, take a few deep breaths and stumble over a sink with the vegetables. I tell Katie to enter an arm wrestling tournament.

She says she's been lifting stuff at the farm since she was a kid. It helped her to be a ferocious soccer goalie for the Elba Lancers when she was in high school.

Tom Rivers and Guilleminia wash zucchini before it will be boxed and distributed to customers the following day.

On to the peppers

I help wash zucchini and squash for a half-hour before Tony, the crew leader of the group I'm in, says it's time to go get green sweet peppers. The peppers are ready for the first time this year. Mike Porter tells Tony he wants 1,200, two for each CSA member.

The peppers grow on plants about 2 feet high. Each plant has about three or four peppers that are ready, and there are plenty of blossoms for peppers that will emerge in the coming weeks.

We have to again bend way over to snap the peppers by their stems. We don't use knives, but we need determined fingers to get the job done.

Tony tries to get me to taste a new pepper that will be available in about a month. He says Mexican folks really like it because it's spicy. I pass on it after he says tasting the pepper will require massive gulps of water. I also may be bound to the bathroom the rest of the day, Tony says.

Tony would pitch the pepper to other farm employees and *The Daily News* photographer. But Tony didn't mention to them the pepper was so spicy.

We spend 45 minutes or so in the scorching sun and we cover all of the rows. But we only have 800 peppers. We're short 400 and we have to comb through the rows once again. We see lots of peppers that are ready. At first they looked like they could grow another inch or two, but Tony says they are fine.

A half hour later, we have the 1,200-quota. We drop off the peppers, and head to the fields of lettuce.

Cut it low

We need 1,200 heads or 60 boxes with 20 heads apiece. This job will require the knife and I can feel my adrenaline rising.

These freshly harvested onions were grown without conventional chemical sprays.

Tony gives me a quick tutorial. Basically cut them low, deep into the stalk. If you cut them too high, the leaves will all unravel and the lettuce won't be any good. Because the lettuce is on the ground, there is some major back bending required. I spread my legs wide, stoop over and lift the lettuce leaves up in one hand. Then I cut as low as I can. I follow Tony's lead and place the cut lettuce into rows, where another worker will put 20 in boxes and load them onto the truck.

Tony works one row and I'm attacking the next one. Soon he is past me, appearing to go at twice my speed. I'm not going any faster because I'm mindful of cutting the lettuce low. I stay vigilant about not dismembering some of my fingers. After 20 minutes or so, I grab a box and start putting the lettuce directly into it, rather than leaving it in a lane for another worker.

We're at the lettuce for maybe an hour, and Tony thinks we have enough. We drive it back to the main barn, where it will be cleaned. I'm sitting on the back of the pickup, with my legs flopped over the tailgate. I don't have a seat belt and there's nothing really to hang on to. I say a little prayer and we make it back safe and sound.

Bag and bean woes

Once we're back, Tony and I help get the bags ready for Saturday, the next day when CSA members get their produce. After using plastic bags in the past, Mike this year opted to try heavy-duty paper bags. But those bags weren't quite strong enough. The rinsed produce, especially the lettuce, would leak in the bags and they would then break down. So the paper bags are now put inside plastic bags. Mike says the arrangement has doubled his bag costs and made for a time-consuming chore. He'll try another approach when he uses all of the paper bags.

Tony works on the bags for about 20 minutes, but he's ready for a new task. He asks me to join him in the bean field not far from the lettuce. We find Billy there, sitting cross legged. He seems very bored.

"Well, you must have done something wrong to get sent to the bean field," Billy tells me. He declares it an awful job. "I'm here because the boss told me to be here," he says.

Tony and I join him. We squat down, but I give up that position after 10 minutes, joining Billy in the cross-legged approach. My knees can only take so much.

The beans are hidden in what seems like a field of tall grass. They are everywhere. After 20 minutes, we advance maybe five feet in the field. It's slow and awfully tedious work. Tony declares he can't take anymore and he leads me back to the main barn.

"This is the one job I don't like," he tells me.

We're back in the barn and there are still at least 400 bags that need to be readied. Tony and I join Katie with the stacks of bags. It takes about 45 minutes to get all 600 ready. It's 3:45, and I'm ready to go home.

I'll be back the next morning, by 5:30, when a group of about a dozen people will put about 7,000 pounds of vegetables into 600 bags.

Before I leave, I tell Mike I liked the challenge of getting the squash, peppers and lettuce. But the beans were pure drudgery. Mike agrees, saying they are one of the more unpleasant jobs.

"Wait until we do the tomato berries next week," he says. "No one likes that either."

I head home. I pass four workers, still at it in the bean field.

Chapter 9

Skipping the middle man: Fresh food, straight to customers

These bags of produce, 86 in all, will go to groups from St. Joséph's Catholic Church in Buffalo. Porter Farms donates one free bag to the church's food pantry for every 15 bags purchased through the Community Supported Agriculture program. (Photo by Tom Rivers)

ELBA — Mike Porter is pacing again at 7 on a Saturday morning in early August. He is worried about his numbers, whether he has enough squash, zucchini, onions, peppers, green beans, beets and lettuce.

The stash of produce, about 7,000 pounds in all, needs to be divvied up equally among 600 customers at Porter Farms.

By 7:30, Porter's crew of about a dozen employees is getting close to packing all of the bags. Mike fears he might be short some lettuce and beans. "How many beans, Tony?" Mike asks one of his workers.

Tony believes the farm, which packed 45 boxes with 20 pounds apiece of beans, is in good shape for that vegetable. But Tony, a native of Puerto Rico who has worked 20 years for Porter, is concerned there won't be enough lettuce.

He is right. Porter is short about 25 heads, and Tony is dispatched to get some more — right away. Tony takes a knife and goes across Edgerton Road in the pouring rain. He is back in about 10 minutes with the lettuce.

Porter is the only farm in Genesee County that uses a Community Supported Agriculture model, where customers pay a subscription for weekly bags of produce. It costs $340 a year for 22 bags spread from June to late November. New customers are added weekly, and they get a pro-rated fee of about $15 a week.

Making sure customers get a variety of produce and enough for everybody has become a greater challenge for Porter Farms, which has seen its CSA membership nearly double in the past two years to 575 members. The farm also gives out some free guest bags each week.

It was easier to plan for 300 customers than 600, Mike says. But he insists the farm can handle more. This year, when he was planning his crops, he made allowances for a CSA that could have up to 700 members. More than half of Porter's business remains with the wholesale market, but Mike would happily sell more through the CSA, where the customers aren't trying to talk him down on the price.

Early morning

I arrive at the farm Saturday at 5:40 in the morning to help fill the bags. The public starts to arrive at about 8:30, looking for the week's assortment of fruits and vegetables. I'm 10 minutes late despite getting up at 4:40, perhaps the earliest I've ever been up. Another straggler would follow me at about 6:30.

The majority of the crew is on time and ready by 5:35, Mike tells me. Four people stay on one side of a series of tables, hauling boxes of the various fruits and vegetables. About eight of us are on the other side, filling the bags. Mike has carefully planned specific counts for each bag. He wants four larger zucchinis, maybe 8 or 9 inches long, in each bag. Two smaller ones, maybe 6 inches or so, also need to be added. As we move along the table we include two yellow squash, two sunburst squash, two handfuls of green beans, two onions, about seven golden beets that are already grouped together with a rubber band, and two heads of lettuce.

I'm one of those chatty morning people — something that annoys my wife — and I'm just talking away while everyone else seems to work in a daze. I have sore legs and shoulders from the previous day's efforts to pick the produce. But I'm in a talkative mood.

I find two "new" people I didn't meet yesterday, and I bombard them with all sorts of questions and commentary. Richard Beatty, 57, and Raeann Engler, 42, of Batavia joined the CSA last year. They started volunteering in the fall when the farm needed help with the Saturday packings.

Beatty and Engler get up at 5 for their Saturday morning shifts. They each work to get the bags open and ready for the workers to fill. A thick paper bag is put inside a thin plastic bag. The bags "can fight you," Engler says, and it speeds up the packing to have the bags ready.

"There's a real sense of community here," Beatty says. "We know the workers and the Porters, and this farm is practically in our backyard."

'Greatest stuff'

The first set of 200 bags of produce is loaded onto a Ryder truck that is driven to stops in Rochester. Most of the other bags will be picked up later in the morning at the farm. There are nine groups from Buffalo, and each week different members take turns and pick up the bags for the group, which range in size from about a dozen people to about 30.

Peggy Grenauer, a former Java resident who now lives in Clarence, is among the first CSA members to pull up to the farm on Saturday at 8:30. A neighbor invited her to join the CSA about four years ago. Grenauer says she is eating much better since getting the Porter produce, and she has a better appreciation for farming.

She likes to learn to cook with some of the more unusual produce at Porter's, including Swiss chard and sunburst squash. Porter issues a weekly newsletter, telling CSA members about the farm practices for the week, and also includes a recipe for some of the fruits and vegetables. Grenauer gets other ideas on www.cooks.com.

"I've learned new recipes," she says. "Or I will take the tomatoes, Swiss chards and peppers and make spaghetti sauce. It's the greatest stuff you ever had."

Eric Gravert, 31, of Lancaster and his wife Jenn joined the CSA this year. They made their first visit to the farm Saturday, stopping along the road to look at the farm's 500 acres of vegetable and fruit fields. They embraced the idea of buying locally grown, organic produce.

"We know it's local and we know it's fresh," Gravert says. "The stuff you get at the grocery store, you don't know where it's come from or what's been done to it."

Veggie delight

While Porter needs extra lettuce Saturday, and has a close call with the beans, there are plenty of extra onions, squash, zucchini and beets. The CSA members are welcome to take the extras.

Kathy Betters of Batavia fills up another bag with the leftovers. She has lots of family in town this week for her son Travis's wedding. She wants everyone to eat healthy before the big splurge on Saturday.

Betters, 58, joined the CSA five years ago. She says it's a good deal, and she feels better since adopting a high-veggie diet. She has tried kale, heirloom tomatoes and other "more unusual vegetables that I never would have tried in a grocery store."

'Sustainable agriculture'

Katie Porter, Mike's 22-year-old niece, manages the CSA and the pickups on Saturday. She gets many phone calls from people outside the area looking for directions or telling her they are running a little late.

When CSA members arrive, she helps load their vehicles. For many of the Saturdays this summer she will be joined by a college intern, Brad Kujawski of Batavia. Kujawski, 22, is wrapping up his studies at Geneseo State College, where he is majoring in history with a minor in environmental studies.

He was at Porter's at 5:30 Saturday, helping to fill the bags. He stayed until about 11, helping to load cars and answer questions.

"This is sustainable agriculture in practice," he says about the farm with its organic methods and community connections. "I like getting to know where my food comes from. You're meeting the farmer face to face. There isn't a 3,000-mile truck drive."

'Strong stomach'

It's an impressive spectacle, counting and coordinating nine types of fruits and vegetables for 600 customers. Mike says the lettuce shortage is only a minor setback and is easy to overcome.

The previous week the farm harvested lettuce, and it looked fine in the field. But later on Friday, the day before the pack, the farm realized many of the heads were moldy inside from all of the rain.

The employees were sent back out to get about 1,000 fresh heads of lettuce that weren't damaged. The workers didn't leave until 8:30 that night, and they were all due back at 5:30 the next morning.

Katie Porter, left, joins workers at Porter Farms on an early Saturday morning in filling 600 bags with produce.

"You have to have a tough skin and a strong stomach to be in farming," Mike tells me. "Because you never know what will blow up on you."

After the crowd leaves by 11 a.m., Mike and the employees plant more beets. Then Mike would bale hay.

On Monday, the workers would begin assembling the 7,000-pound produce collection for the following week. Katie tells the CSA members they can expect berry tomatoes and cucumbers in the next bags.

Chapter 10

Cutting cabbage proves hardest harvest yet

King Kong, left, gives a head of bravo cabbage a gentle toss into a bin while working with other Jamaicans in harvesting cabbage on a muddy day. King Kong, whose full name is Daniel Dinnal, has been working at a cabbage farm for two years in the Western New York. (Photos by Nick Serrata)

LYNDONVILLE — The morning of the ultimate endurance test, the day I would try to cut cabbage, I was certain the challenge would be delayed because of a hard-driving rain, thunder and lightning, and deep mud in the fields.

But I would learn there is no weather too awful to cut cabbage, especially when a farm has to have 40 tons picked that day and shipped to New York City the next morning.

On a Monday in mid-August I joined eight men from Jamaica, chopping cabbage and heaving it into 1-ton bins. I would last nearly 10 hours with them, going until I could no longer squeeze the knife to slash off a 20-pound head of "mega-ton" cabbage.

It was a wild day, and I'm not sure I can adequately describe the misery and exhilaration I felt, standing in at least a foot of mud, plunging a knife through a head of cabbage and then flinging the vegetable into a bin. When the day ended for me, I could barely grip the steering wheel of my car. The tendons near my wrists were swollen and my pants were caked in mud.

Rare local man sighting

I show up at Lynn-ette and Sons in Kent a little after 7 in the midst of another downpour in this summer of rain. Darren Roberts, the farm's co-owner, agrees to let me try cutting cabbage. Among his many tasks, he oversees the farm's 700-acre cabbage operation.

Darren, 42, seems a little bewildered by my interest in the job. He can't remember anyone from around here ever wanting to cut cabbage. The farm puts many ads in local newspapers seeking field help, and no locals have even called about a job in at least two decades, Darren says.

I tell him I want to do more than try cutting cabbage. I want to see if I can do it for eight hours. He can't contain his chuckle. He asks if I have boots. I don't have the long rubber boots that go up to the knee. But I have some that are about a foot high, about halfway up my shin. I figure they'll do the trick. I tell Darren I'm eager to get going, despite the downpour.

Darren and one of the tractor operators, Spencer Heidemann, hop into a pickup truck. I follow them in my Tercel to a field about 10 miles away on Platten Road outside Lyndonville. The rain is even more intense when we get there, with swirling dark clouds and thunder.

Big order

Darren says the farm has to have 40 tons of bravo cabbage loaded by the end of the day and in New York City by morning. There are also orders for "mega-ton" cabbage due for kraut processing plants in Geneva and Shortsville. The 8- to 10-pound heads of bravo cabbage need to be treated with care — "Handle them like eggs," Darren says because they are going for the fresh market and won't just be run through a processing plant.

I pull on my boots and we walk about 50 yards across a ridiculously muddy field. We maneuver across some deep ridges from where the tractor tires ran across the field. We pass rows of harvested cabbage. The heads are gone and only giant leaves remain lying in the mud.

I tell Darren the cabbage are "serious plants." They have somehow persevered and thrived through steamy-hot weather and tons of rain.

There's a tractor holding four 1-ton bins made of heavy-duty cardboard, with two bins at the front of the tractor and two behind it. The Jamaicans, all wearing baggy raincoats and rubber pants, speedily slash at the cabbage and fling it into the bins. Darren tells them to go easy with the cabbage and only throw it if they are within a few feet of

Bob trims a head of cabbage and sends rain drops flying. King Kong, back left, looks for cabbage to catch and set in big boxes.

the bins.

He introduces me to the crew, tells them I'm a newspaper reporter trying to learn about farmwork. A guy named Bob steps forward and shakes my hand. "It's an honor to have you with us today," Bob says.

I find out he's 50, the oldest guy in the crew. He has six kids and has been working at American farms for almost 20 years. This is his second year with Lynn-ette and he's one of 100 Jamaicans at the farm through the federal H-2A program that allows American farms to bring in legal foreign workers on a temporary basis.

Lynn-ette lost most of its work force two years ago in immigration raids. The past two years the farm has brought in 100 H-2A workers from Jamaica, plus 60 from Mexico. The workers are paid at least $9.70 an hour and the farm must supply housing and transportation from their home countries and then back to either Mexico or Jamaica. Some workers stay six months and some like Bob will stay for 10 months.

Robin Roberts, Darren's brother and a farm co-owner, says the workers are dedicated and do a great job, helping the farm deliver top-quality produce. And he doesn't have to worry about the sudden removal of his crew.

No 'hacking'

I watch Darren chop the cabbage, a ferocious deed. He seems unfazed by the rain dripping off his hat, the thunder in the background and the loose footing. After I watch him for about 5 minutes, he hands me what looks like a big putty knife. It doesn't have a pointy tip like the knives in my kitchen. The cabbage knife's blade wraps around in a rounded tip. Darren tells me to just turn the cabbage on its side and make a quick plunge with the knife.

I lean a cabbage head over and try to slice it from the side of the knife. I don't quite separate the head from the stalk and make a few more chops.

"Don't hack at it," Darren advises. "That's how some of our guys have needed stitches."

Darren again demonstrates the tip-the-head-and-plunge method. I try again and unleash some fury, managing to make a clean cut in one motion. I get a decent string going, of cutting off the cabbage heads, grabbing them with both hands, and tossing them in the bin.

Darren seems satisfied I'm not going to maim myself or other workers in close proximity, or destroy his vegetables. He has other things to do and leaves the work crew.

We're bunched up a bit, the nine of us, removing cabbage from six rows. I'm a little nervous about slashing the guy next to me, only a few inches away. And I'm fearful of gouging someone with the knife when I toss the cabbage. I usually throw it with my right hand, the same one I'm using to carry the knife.

Bob comes over and suggests that he will cut the cabbage if I throw it in the bin. He cuts for me and the guy next door. That eases some of the crowding near the bins. Bob and I switch after 15 minutes or so. I find I'm cutting with a vengeance when I only have to focus on that task. But several times, when I cut loose a cabbage head, muddy water is unleashed right in my face. A few chunks of mud make direct hits on my eyeballs. I try to squint it out. Despite the discomfort, I keep cutting.

A powerful rain storm leaves this field of cabbage flooded. The workers from Jamaica still manage to get the crop out of the field.

Water, water everywhere

I'm distracted by all the water and mud filling my boots. At a foot high, the boots are far too short for the deep puddles and mud out in the fields. There's no chance of keeping it out of my boots. The other guys have higher boots with rubber pants.

About every half hour, I lean against the tractor and empty my boots of mud and water. My white socks have turned dark brown. I get a lot of laughs from the workers. Bob has mercy on me. He thinks I should get my pant legs to cover the boots. He even helps get my right pant leg over the boot while I wrestle the left pant leg over top. My jeans are no good as a water repellent, however. The boots quickly fill again.

King Kong has to fight to get his feet out of the deep mud. The workers wouldn't be stopped by the loose footing.

It takes maybe 10 minutes to fill a 1-ton bin. There's a brief stop in the action so the flaps on the box can be taped to stand up higher. The Jamaicans stand the four flaps up on a box, and run circles of black tape around the box, maybe five or six times. One guy usually stands near the box, and catches some of the cabbage heads, gently setting them in the box so the produce doesn't get bruised.

It keeps raining and lightning enters into the mix. We see it strike at least three times, and once it hits about a half-mile away. No one else seems troubled by the fireworks.

"This isn't so bad," says a guy named Daniel, who is called King Kong by the other workers. "You should see it in November when it snows. That's bad."

King Kong is 33 and is working his second 10-month stint for Lynnette, arriving in January and leaving in November. He and about a dozen other Jamaicans will trim storage cabbage the first three months they're here. Then they will plant crops in the spring. They will be harvesting the vegetables for the next three months.

King Kong is 6 feet 3 inches and weighs 260 pounds. He seems well-liked for his cheery disposition — he frequently bursts into songs about God or about "Joshua." I'm not sure who Joshua is. I'll find out later.

King Kong seems worried about me. He tells me I'm not cutting the cabbage quite right. I need to make a quick, clean cut rather than my style, which sometimes takes two cuts or more. King Kong says my right hand won't last with my inefficient approach. He also tells me I need to get some rubber gloves on or else my skin is going to fall off.

I have puffy bathtub-skin already, and I can see two open spots emerging on my hand. King Kong offers to chop the cabbage for me if I throw it in the bins.

Time stands still

I'm beginning to feel mighty thirsty and hungry. I tell King Kong it must be close to lunch time. "That's midday, in about two hours," he tells me.

I can't believe these guys don't eat until 2 p.m. because surely it must be about noon. I see Spencer, one of two tractor operators, and I ask him the time.

"It's 10:16," he says. I'm a person who generally doesn't swear. But I utter a few choice words. Two more hours until lunch? Seven more until this is over?

But I don't have to wait long for a break. The workers need to get more boxes ready so we leave the cabbage and go to a spot next to the field where there's massive folded cardboard. The guys start shaping it into boxes, even using chainsaws for part of the job.

I need a drink, and immediately gulp down a quart of water. I gobble two granola bars. I yank off my boots and peel off my socks. I root around in my car and come up with a 1996 Garlic Festival T-shirt, which I shove down my boots to soak up some of the mud. Deeming my socks worthless, I decide to go barefoot in my boots for the rest of the day.

Tom Rivers heaves a 10-pound head of cabbage into a big cardboard box.

There's a crew leader-type worker there named Miguel. When he's got all the boxes ready I ask him about a pair of gloves. He finds me a pair of bright orange ones. I inquire about some longer boots, but he doesn't have any.

Swallowed by mud

We're back in the fields. Even though the rain has stopped, it seems muddier than ever. We hit some real deep spots, and I can barely walk. The guys tell me to walk on the plants and stay out of the rows of mud. That's not always possible, with all the foot shifting needed to chop, grab and throw the cabbage.

I'm getting some laughs and pity because several times I have to really fight to get my feet moving in the mud. There's the added pressure of a slow-moving tractor coming toward me. One time I just can't get my legs moving in the thick mud. The tractor is closing in and I decide to just hop out of the boots. I'm barefoot in the mud. I jerk my boots with both hands and crawl through the mud out of harm's way. A worker named Winston is driving the tractor. He toots the horn and gives me a "thumb's up" for my escape.

We keep cutting, filling the bins, and my right hand, although in much better shape with the rubber gloves, is struggling to unleash the needed fury on the cabbage stalks. I'm not sure what arthritis feels like, but my joints, tendons and everything in my right hand just aches.

We finish our 18th bin. King Kong says it's noon and time for lunch. We trudge the hundred yards through the mud to get off the field. We're all ready for a half hour of peace. But Miguel says he needs four more bins. Miguel has 18 bins loaded on a tractor trailer, but there's room for four more. He doesn't want to waste space on the truck.

King Kong leads us back. I don't sense much grumbling. I guzzle some more water and feel slightly revived, although I still haven't got used to the mud oozing between my toes or the boots chafing my shins. The long walks don't help.

Lunch, finally

We fill the four bins in about 40 minutes, and finally, I get a chance to sit down. I see Spencer, the 18-year-old tractor operator, in a van, eating his lunch. I have a few questions about the bins and farm operation. I camp next to him in the van and I'm distracted by his two Mountain Dews. I offer $10 for one, and he gives it to me for free.

Spencer graduated from Kendall last June. He grew up working on a farm owned by his father Scott Heidemann. That farm isn't big enough to support another full-time employee so Spencer has joined Lynn-ette. He says today's rain was "pretty bad."

The past few years in early August the workers have been out in 90-degree heat and the farm usually has to irrigate because it's so dry. This year, Lynn-ette hasn't irrigated at all.

Spencer has some prep to do with the bins and tractors, so he leaves.

I go join the Jamaicans in their bus. Winston, the tractor operator, has an international driver's license and he drives the workers back and forth from a Ransomville labor camp. It's an hour-drive each way.

'A means to an end'

King Kong is sitting in the driver's seat. He's done with lunch. I plop down in the seat behind him. He tells me he worked in landscaping, furniture making and pool cleaning in Jamaica. But the jobs were never steady year-round.

He's married with five kids, ages 2 to 16. He agreed to try farm work last year for the first time because of the steady income for 10 months. He has an international calling card and speaks with his family each day. "It's expensive but it's something I need to do," he says.

Spencer Heidemann, a tractor operator for Lynn-Ette and Sons, tapes the top flaps of a box to fit in more cabbage.

King Kong applied to be in the H-2A program for two years before he was approved by the Jamaican government. He said the job will help him send his children to good schools.

"It's hard being away from them but it's a means to an end," he tells me. "I could be there and not support them, or I can be here and get them the things they need."

King Kong , whose full name is Daniel Dinnal, says all of the workers come to give their families better opportunities at home. He and the other guys in his crew are up at about 4 each morning, preparing their breakfast. They leave for the farm at 6 and often don't return until 8 or 9 p.m. They get Sundays off.

"I came here with an open mind," King Kong says. "I had no expectations because I had never done this type of work before."

He says he's grateful for the chance to work for Lynn-ette. "I have to be mindful of who I'm doing it for," he says, calling his children his inspiration.

Heavy, heavy cabbage

After lunch, we move across the road to another field of cabbage. This stuff is much bigger than the bravo cabbage. Darren joins us in the effort.

He tells the workers the "mega-ton" cabbage will be used by a kraut processor and doesn't have to be handled with as much care. Just chop and toss it in the bin. It doesn't have to be handled like a carton of eggs, he says.

One of the workers from Jamaica trims cabbage while more cabbage is airborne. One of the Jamaican workers lost 30 pounds his first three months on the job.

This mega-ton cabbage feels like a heavy bowling ball. I cut it and throw it, but it takes a mighty effort because of the weight. I suggest to Darren the Buffalo Bills come out and heave this cabbage as part of their training camp.

Because this cabbage is so big, we can fill two bins in about five minutes. Then the tractor turns around and the other two bins get filled. It's 10 minutes of high-paced chopping and tossing and then we get about a five-minute break before the next tractor is ready.

We all sit on the enormous heads of cabbage while we wait. I find out the workers are ages 27 to 50. At 34, I'm one of the younger guys in the group.

Joshua revealed

Someone is again singing "Come on, Joshua" and I ask the guy next to me if he's Joshua. Nope. His name is Desmond. He tells me Joshua is the Jamaicans' name for the sun. When it rains, they sing for the sun to come out.

Now they're singing because they're happy with the weather. The sun is in its glory and everyone left their rain coats off after lunch. I can now see all of their faces and I can see their big arms and shoulders. Bob, the 50-year-old, looks to be the toughest dude in the group. He's built like a boxer.

Workers trim some cabbage in close quarters. The farm, Lynn-Ette and Sons, needed 40 tons of cabbage in New York City the next day.

Desmond tells me the workers really crank up the "Joshua" songs in October and November, when they're out harvesting cabbage in the snow. Desmond is 38 and he joined Lynn-ette in May. He's already dropped 30 pounds in three months.

We continue to fill the bins, working 10 minutes in a frenzy and then getting a five-minute break. It's around 2:30 p.m. and King Kong tells me he can't believe I'm still here. The guys figured I would have left hours ago.

I tell them all I'm trying to get in my eight hours. I tell them they're fun to work with but I confess my right hand can barely grip the knife anymore and the boots have rubbed the skin and hair off my shins.

Desmond says he'll cut the cabbage if I toss it and that's the plan for awhile. I sneak in a few cuts but I think my fingers are dead, which makes it hard to grip the 20-pound cabbage and heave it.

I try cutting with my left hand and miss the stalk and nearly jab one guy in the boot. I'm mainly just a thrower now, and I'm using my left hand for most of the work.

We continue moving around the field, hitting different cabbage patches. There's one low-lying spot that is flooded with 400 or 500 heads of cabbage. They all have degrees of rot on the leaves. Some of them have an awful stench.

We spend 15 minutes inspecting and trimming the heads, trying to determine whether they're salvageable. We can cut some of the rotted leaves off and the cabbage could pass the processor's muster. But Darren decides he doesn't want to risk it. The processor could reject all the cabbage based on the iffy ones from this spot. So we move on. We're not short on cabbage options.

8 isn't enough

It's 3:30 p.m. and I've hit my eight-hour goal. Darren gives me a ride in the tractor back to my car. I get a drink and decide I'm not done after all. I'd like to finish the shift. Maybe these guys will be done at 5. Then I could interview them and get their last names. Most of the information for this story comes from shooting the breeze in the field. I don't have my pad and paper in hand.

Darren finishes unloading the bins and gives me a ride back to the workers. I have a new burst of vigor in my right hand and for the next half hour I'm cutting, throwing and doing the full job.

Not long after 4 I give up trying to cut anymore with the knife. I just can't seem to grip it anymore. It's a struggle to grab the cabbage but I can still do it. I can't throw it from 15 feet anymore. I try to stay close to the bins and I alternate my throwing motions, from the underhand toss, to shooting it like a basketball, to some overhand, using my left and right hands.

Bob gives me high praise. "You're doing good, man," he tells me. "You're keeping up with us."

Bob's words are a boost, but I soon feel deflated. King Kong tells me the workers will be there until 7 or 8. He doesn't seem to mind. After they finish the mega-ton cabbage, they have 18 more tons of the smaller bravo cabbage to harvest.

It's getting close to 5 and I tell the guys I'm just going to finish this last batch of mega-ton cabbage. I can no longer throw it, and I'm feeling shooting pains just trying to grab the stuff. I just walk it to the bins and drop it in. I'm barely helping the cause at this point, but I want to do what I can. I get through the next 10 minutes and tell everyone goodbye. I get a few handshakes but I think they can tell my hand has moved into the throbbing stage. Most just wave to me.

Winston gives me a ride back. He's 38 and worked as a truck driver in Jamaica. He says he can make more working for Lynn-ette in five months than he can the whole year in Jamaica. He has a baby at home and he is focused on a goal for his family.

"Some people buys cars and other stuff (with their money from the farm)," he says. "But I'm saving up for a house."

He invites me to visit him someday in his home. I tell him that would be an honor. I manage to give him a hearty handshake.

Chapter 11

Scrambling to catch up in the cuke rows

Workers at Torrey Farms hunt for cucumbers and fill baskets that hold 30 pounds of produce. A fast worker can fill and unload a basket every three minutes. (Photos by Nick Serrata)

ELBA — On a Monday morning in late August, I headed for the fields of Torrey Farms, intent on keeping pace with the Mexican cucumber pickers. The day didn't go quite as planned.

Maybe it was all the Olympic hype and glory I had just witnessed that made me think I could achieve a gold medal effort among the cucumber crew. But the job nearly crushed my spirit after a mere three hours.

I told Maureen Torrey Marshall, the farm's co-owner, that I was certain I could get through the shift. I had recently survived 10 hours in a cabbage field during a massive downpour amid deep mud.

Two weeks ago I wrote that cutting cabbage was a miserable job: "The hardest to harvest," the headline said in the Aug. 16, 2008 *Daily News*. Some of our local growers told me cukes are even more grueling to pick than cabbage. Maureen was "kind" enough to invite me to join a crew of 23 other guys in a cucumber field. I would even be "chipped" so I could see how my output compared to the other workers.

I was curious and overly confident in my skills and stamina. I figured my past farm jobs – since late April – had adequately acclimated me to the rigors of farm work. I've dropped 25 pounds and added a little bulk. Bring it on.

Showtime

I show up Monday around 7 in the morning to join Torrey's workers at a 30-acre cucumber field on Ridge Road in Elba. A former school bus, with thick blue and white stripes painted on the sides, is parked by the road. I know the work crew is somewhere nearby. I don't see them, but I can hear a tractor in the distance.

I round a corner leading to the cucumber field and see two crews picking cukes. They each have about two dozen people, with about 12 people on each side of a tractor. The workers — all men — are bent over, hunting for cucumbers. They put the cukes in buckets and carry the pails on their shoulders. They high-step the vines and make it to a wagon behind the tractors. The workers push the buckets as high as their arms will go, and dump the cucumbers in the wagon. Then they scurry back to their rows in the field. All the pickers look very busy.

Joe Menas, the field superintendent, is quick to greet me and give me a crash course in cucumber picking. Don't pick small ones shorter than 6 inches. Don't pick big fat ones that look like long watermelons, and stay away from the ones that are crooked. Also pass on ones with lots of scratches and scabs. The ones with rot, obviously skip.

Joe also warns me about the prickly vines and stiff hairs on the stems of the cucumbers. Even the leaves can scratch you, he tells me. I'm glad I remembered my leather gloves. I forgot my boots, though, and I'm wearing running shoes.

Joe hands me a heavy-duty plastic bucket that holds 5/8 bushel of

A crew of about two dozen workers picks cucumbers for Torrey Farms. It takes the farm about 10 weeks to pick 1,000 acres of cucumbers. The fields get picked two or three times.

cukes, about 30 pounds when loaded. He shows me my row and I get to it. No knife is needed. Just grab the cucumbers and twist them off their stems. Some I can do with one hand, but I usually use both hands, one on the cucumber and one on the stem.

This involves serious bending and these cucumbers don't sit in a straight line. I brush aside leaves and vines and feel very annoyed when I find cucumbers with rot spots. They're no good.

Many are in fine shape and I fill my basket and head for the wagon, nearly tripping over some of the vines. I make the 10-yard walk to the wagon and one of the workers reaches down to grab my pail. He dumps it and I turn around and he touches a probe against the chip I attached to the back of my hat. I'm officially in the Torrey Farms database.

Tom Rivers races to get an overflowing basket of cucumbers to a wagon. Rivers spent much of the day lagging behind the Mexican crew.

Lost

I head back to my spot, but I'm not sure where I left off. The field of vines and leaves all looks pretty much the same. I find a row and start picking again. Joe swings by and tells me I'm two rows too far. So I move over.

I'm back to bending, hunting for cukes. This is the third time Torrey workers have picked in the field, so it's not as plentiful as the first two rounds. Joe says it's easier to lose your spot today because the vines have all been stepped on. With the first pick, when the vines and leaves were robust, it was obvious where you left off, Joe says. But now it's a little trickier to find your way. He says some pickers make a cross with two cukes to show where they were. If it's muddy and the ground is soft, pickers will stick a cucumber in the ground as a marker.

Today the ground is hard and I find the leaves cover up any attempts to make a cucumber cross. I just assume I can find my spot. I'm about five rows from the wagon. But I keep coming back to the wrong row, and Joe keeps having to correct me. I'm falling behind the other pickers and Joe joins in my picking effort, helping me fill my bucket.

He also tells the guys at the wagon to stop helping me dump the cucumbers. He thinks I can heave the 30-pound bucket over the rail without a problem. It takes a few tries, but I find I can do it without causing cucumbers to topple down on my head.

Joe has another crew to check on, so he leaves me after an hour or so. I try to stay near a guy named José. I find out he's 20. He tells me not to carry the heavy pail by my side. He tells me to get it up on my shoulder. I try it with the next trip to the wagon and it's a marked improvement. I don't feel any ache in my back like I did with the other carrying method. It's also easier to pop the pail up and dump the cukes into the wagon.

I'm not nearly as fast as José and he often picks some of the cucumbers in my row. Sometimes he sets them in a pile for me to add to my bucket and boost my numbers in the Torrey system.

José offers me another tip. He shows me how to straddle the row, with my legs spread out. Then lean forward and brush through the leaves. The fast pickers can grab cucumbers with their left and right hands. I try this position but it's a lot of strain on my back and glute muscles.

I find myself alternating from bending over with my legs spread out, to squatting, to kneeling. The task gets harder as the basket fills. Once it's half full, I have to drag at least 15 pounds with me.

After about three hours I feel like I've been trampled by wild bulls. My back, legs and wrists are aching and I don't see how I can last until 5 p.m. I'm going a little bit faster than in the beginning. Joe notices. But I'm still the slow guy who can't find his row.

Help

I feel some rejuvenation when we finish one of our passes across the 75-yard field. There's a new beginning on the field next to a 10-yard strip of grass and weeds, ground that wasn't good enough for crops. I immediately take a spot next to the weeds. I know I won't get lost coming back from the wagon. This row also seems loaded with cucumbers. I get into a nice groove and I'm almost keeping up with the Mexicans.

A Torrey Farms employee heaves cucumbers into a wagon, while the other worker prepares to arrange the pile and credit the worker's output with a probe in his right hand.

After 30 minutes a guy named Enrique asks if he can join me in the row. He notices I'm still a little behind. I tell him he can work the row, too, although I'm wondering how this will work because the pickers don't seem to share rows. Enrique goes 10 feet ahead of me and starts picking. Two minutes later he dumps 15 cucumbers in my bucket. I've got a full load so I go to the wagon.

Enrique repeats this many times in the next 45 minutes, dumping cukes into my bucket or leaving a pile of 10 to 15 cucumbers for me. He also manages to fill his own bucket and unload them in the wagon.

Joe tells me this is typical Enrique behavior. He often helps the slow guys, or pickers who may be sore. Enrique, who's 28, does all that and still has some of the best picking stats on Torrey Farms.

Joe Menas, a cucumber picking crew leader for Torrey Farms, keeps Tom Rivers company while he fills a basket. Menas joined the farm after about 8 years in the Marines, including stints in Iraq and Afghanistan.

The help from José and Enrique does more than help me keep up. It lifts my spirits, and I don't feel hopeless. I feel like the crew wants me to get through this mighty task.

From Iraq to Elba

It's lunch time. Joe invites me to eat in his white farm Bronco. The other workers head for the old bus. I've got four bottles of water in my car and two bananas. Joe eyes my selection and offers me a ham or turkey sandwich, whichever I prefer. I go for the ham. He also has chocolate chip cookies and I eat three or four of those.

Joe, 30, is tall and lanky with a blue "Torrey Farms" shirt. Last year he completed 8 1/2 years in the U.S. Marines, including two stints in Iraq and one in Afghanistan. A West Virginia native, he married an Elba girl – the former Kate Gaylord – nine years ago when they met in North Carolina, where he was stationed at Camp Lejeune. He and Kate have three sons, ages 4, 6 and 7. Joe joined Torrey Farms when they moved to Elba last year.

Joe received a Purple Heart after his Humvee was rocked by a roadside bomb in Iraq. He suffered a severe concussion and traumatic brain injury. When he was discharged from the military, his wife wanted to return home. She is studying to work as a paralegal.

"My wife followed me and my military career for eight years," Joe says. "It's her turn. She wanted to come home and be near her family."

In Iraq he led a platoon of 40 soldiers, many of them Hispanic and many not-yet American citizens. Joe says he has high respect for the Latino soldiers and the workers at Torrey Farms. They have an incredible work ethic, he says.

He initially drove a tractor for Torrey's, helping the farm harvest cabbage. He doesn't like the tractor job as much as working in the fields with the workers. It reminds him of his Marine days, leading a group towards a common goal. He refers to the cucumber pickers as his "civilian patrol unit."

This worker shows the proper way to carry a 30-pound basket of cucumbers. Hold it high so there's less torque on the back. A crew of two dozen workers filled 3,213 baskets on this late August day.

Joe leads the cucumber picking for about 10 weeks. When that's done in about a month, he'll shift to cabbage harvesting. Then he'll fix equipment in the winter and help with potato packing. In the spring he will run transplanters that get the onion and cabbage crops in the ground.

"It's a good job," he said about his new career. "You're always doing something different."

Pressing on

After lunch we're ready to start a new pass across the field. I can no longer plant myself next to the row of weeds. So I take the second row by the wagon. I figure that will keep me from getting lost.

My back is a little stiff, and I can plainly see some swelling in my wrists. But I'm feeling patriotic. I can't help but think of Shawn Johnson, the gold medal-winning gymnast, and I decide to press on with as much force as I can muster.

I'm aware that I may be the only American these guys ever see working as a cucumber picker. (Joe, while he's not afraid to get his hands dirty, is a supervisor and not a field worker.) I want to leave a good impression, and represent my country well.

Turns out I still have a lot left in the tank, and I just ignore the aches and pains. I'm next to José again and mostly keep up with him. Joe helps me a little and José grabs some of my cukes to prevent me from falling behind the pack.

Snakes and soda

Joe gets quite excited about every two hours or so when someone finds a garter snake. He insists on picking it up and removing it from the premises. I don't know what he does with it. I know he doesn't like them.

I don't see any snakes on the ground, thankfully. I don't think I would handle that well. But there are many little frogs hopping around.

We are clearly going much faster across the field this pass. There aren't many cucumbers. This section has been picked over twice and new cukes just haven't emerged like in other parts of the field. On our next pass, I purposely pick a spot away from José and Enrique. I don't want any helpers. I'm feeling an early afternoon energy rush and I want to see how fast I can go.

I find a spot next to a fat tire track that runs a good way across the field. That track will help me find my row. I see lots of cucumbers and I quickly fill my bucket. I'm even ahead of the two guys to my left. I get a good groove going for about a half hour. Other pickers must have scarce cucumbers because more people come over to my row. They go ahead of me and I'm left with the cukes they don't want. They're still good cucumbers, just trapped under vines.

My brisk basket-filling pace slows with the added pickers in my row. We polish off the pass and there's nothing left to pick. Joe says we'll go to a different field about a mile away on Graham Road. I get in my car while the other workers head over in the bus.

The other field is smaller and has several big spots without any plants. It looks like the barren land was flooded and too muddy for the crop to take root. The field is only about 30 yards long. It's maybe 3 in the afternoon and I notice several of the guys have lost their pep. I make a point of trying to pass them. That seems to light a fire in them and they quickly speed up, leaving me in the dust.

Joe says he's going on a soda run and he gets orders from everybody for Pepsi, Sierra Mist, Root Beer and other flavors.

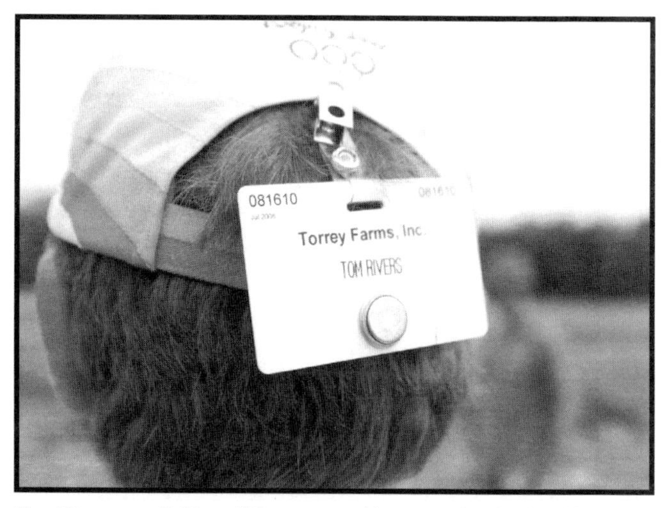

Tom Rivers was "chipped" to measure his output for the day. Torrey Farms has been using the technology for a decade.

Almost lost

The pickers start a new pass and there aren't any clear markings where my row starts or ends. It's another wide-open prairie of cucumber vines and leaves. And I've got back, leg and wrist issues. I have two world-class sprint pickers by me and soon I'm trailing the pack.

I've got lots of cucumbers in my row. Before, I begrudged the help from Joe and Enrique. I didn't like being treated like a charity case. But I wouldn't mind a little help now.

The worker above uses a probe to record the other worker's output. The technology allows the farm to easily keep track of the work for 88 cucumber pickers.

I start to get some, but I don't know where it's coming from. There's a pile of 10 cucumbers already picked waiting for me. I go dump my basket and return to find another 20 cucumbers. I have to fight back tears because I'm so surprised and thankful for these acts of goodwill. Without the help, I doubt I'd survive. I would just fall farther and farther back.

Joe returns with the sodas and to a hero's welcome. Some of the workers give him a buck or two, but he pays for many drinks himself. He insists I take mine for free. Joe has soda for the second work crew and he passes those out. He also finds another snake.

He returns and helps me try to catch up. He thinks Hector, the crew chief, is the mystery angel leaving cucumber piles. But it appears other guys also are chipping in with the effort. With all the help, I eventually catch up with about half of the crew.

There are eight to 10 guys who are star pickers and they seem to have their own race to get across the field first.

Almost done

It's a little after 4, and Joe says we'll call it a day around 4:30 when the wagon is loaded. He doesn't want to start another load that would take until 6 p.m. to fill with 22,000 to 24,000 pounds of cukes.

I'm pleased the end is near. I'm covered in dirt. Every time I unload the basket of cucumbers, soil rains down on my head.

Joe and Hector both are helping me keep up. Despite my throbbing wrists, I still manage to heave the 30-pound baskets of cukes on my shoulder and then dump the produce in the wagon.

It's about 4:30 and the wagon is full. But Joe decides we will go another half-hour. Hector convinced him. The workers don't want to stop. They want to go until 5 so they get a full 10 hours of pay. They started at 6:30 and had a half-hour lunch.

That extra half hour is bad news for me and some of the other guys. I see one picker just shuffling through the field, not even trying to bend over for a cucumber. That fires me up a little, but I'm not going with much zest, either. I can't bend over anymore. I can only squat down to get the cukes.

Hector and Joe throw a few cucumbers in my bin. But Joe gets distracted by a mouse carrying two babies. He chases after it and finds the babies but the mother is gone. He says he will take them home and try to feed them with milk through an eye-dropper. He has been nurturing a baby rabbit he caught on Saturday. "I just love animals – everything but snakes and scorpions," Joe says.

I push myself to 5 o'clock. I spend the last 10 minutes helping to fill other guys' baskets, adding a few here and there for the other pickers. I don't want to try for another basket of my own. I don't think I can carry 30 pounds.

How I rank

At 5, Joe calls it a day and we all make a slow walk across the field. I drive back to Torrey Farms' main office on Maltby Road. Joe says I can see a breakdown of the workers' output for the day. Maureen has the data.

We find out the crew picked 3,213 baskets in 212 manhours for an average of 15.15 baskets per hour. José Mayo was the star of the day, picking 202 baskets for a 19.24 hourly average, or one about every 3 minutes.

Enrique, the guy who helped me for one pass across the field, finished fourth on the day with a 17.43 average. I see his last name is Lopez.

I came in last, no surprise, but really didn't do too badly. My 10.21 average isn't too far from the slowest Mexican picker who had a 10.79 rate. (This guy has had a sore leg for about a week, Joe tells me.) So I'm only a little slower than the guy hobbling on one leg.

Joe deems my efforts a good showing, better than some of the Mexicans on their first day. Maureen also says I did well. "You finished," she says.

The farm uses the data to determine yields in fields. The numbers also will help the farm and federal government determine productivity standards for cucumber pickers in the H-2A program, which provides temporary work visas for farms. Torrey's has 88 H-2A workers here this year. They each make $9.70 an hour and receive free housing and transportation from Mexico to Elba and then back home after the fall harvest.

Joe thinks they're all good workers. He says they often average at least 20 baskets an hour on the first and second rounds of picking in a field. But with the third round, there are fewer cucumbers to grab, so the numbers are lower.

The pack house

Maureen gives me a tour of the packing operation. After the cucumbers are picked they go to the packing shed next to the main farm office. The cucumbers are washed and then sorted into five groups by length, ranging from about 6 to 15 inches. They all get an edible vegetable wax coating to prevent dehydration and also to make them look shiny.

Maureen says most of the cucumbers picked that day have already been packed and will be "on the road tonight." Some are going to military bases in the South, while others will go to markets and chain stores in Philadelphia, Boston, Michigan and Maryland.

The South prefers the smaller 6-inch cucumber while the Baltimore and Philadelphia markets prefer the large cukes. Most people, however, favor cucumbers that are 8 to 10 inches.

We return to the main office and Maureen proudly shows off two massive soccer trophies. Her workers have twice won the championship in a farmworker soccer league in Brockport.

One team was made up entirely of her cucumber pickers. "They had better endurance," she says.

The following year, the packing house employees, who also work long hours, won the trophy.

Chapter 12

Learning from a public market pro: 'Don't be stingy'

The Rev. Wilfred Moss was at the Public Market in Rochester at 5 a.m., preparing produce for a rush of customers. An estimated 20,000 people visit the market on Saturdays. (Photos by Tom Rivers)

ROCHESTER — I tell him not to go easy on me. I want to learn from the master, a man who is regarded as perhaps Albion's hardest-working entrepreneur.

The Rev. Wilfred Moss, 77, seems happy to oblige. On this early September Saturday, for about eight hours during a hectic day at the Rochester Public Market, the Rev. Moss has me fetching heavy bags of onions, potatoes, ginger, beets and other produce from the back of his white truck.

He instructs me on how to present the product, how to speak to customers, how to handle money, and how to keep the trays and baskets restocked.

"Don't be stingy and don't give them too much," he says about a selection of 20 different vegetables. "We have to make a profit."

He chastises me for the rare idle moments when I'm just standing there eyeing the crowd. "Look busy," he says

He doesn't like my occasional joking around with customers. "This is serious business," he says, and gives me a stern look.

Moss has nine grown children, and three who live locally — daughters Cheryl, Naomi and Ruth — are working with us at the market. They tell their dad to give me a break. "He asked me to show him how to work, and I'm showing him," he responds.

Moss has been a vendor at the Public Market in Rochester for 35 years. He often goes Tuesday, Thursday and Saturday each week throughout the year. He doesn't miss Saturdays, the busiest day when more than 20,000 people can be expected from 6 a.m. to 3 p.m.

Moss clearly loves the people, and they love him. Many shake his hand and hug him, calling him "Rev." He calls them "My brother" or "My sister" and ends most conversations with "God bless."

The Moss motto

Moss has been a longtime family friend. He lives on Washington Street in Albion, just around the corner from my in-laws. His kids grew up with my wife and her siblings. In that neighborhood, Moss is clearly a respected resident.

He started a barbershop next to his home in 1960. Blacks couldn't get served by the other shops, so Moss started his own. When blacks were rejected for housing, Moss started buying property, improving it, and renting and selling to them.

He has been a force for social justice in Albion, that I knew. I heard he had some farm venture, but I didn't know the extent until a Saturday in mid-September.

If you're a vendor at the Rochester Public Market, it's nothing to take lightly. There are tons of produce to haul and prepare for the market. Moss grows a good share of it at a farm in Albion's countryside. He also has friends in the business, and he will sell their onions, garlic, ginger, potatoes and other crops.

"I just pick up the phone," he says. "It's not too hard. The mind of a person makes that person. That's my motto."

A full schedule

Moss grew up in the Bahamas and came to the United States when he was 21, on Nov. 12, 1952. It's a key date in his life and the Rev. says it with emphasis: "November twelfth, nineteen fifty-two."

He was on government contract to work at citrus farms in Deerfield Beach in Florida. He emphasizes the location, too. Not just Florida, but Deerfield Beach. He picked apples in Virginia and worked for the Green Giant company in Minnesota. Then he came to Albion in the mid-1950s.

Cheryl Moss has a variety of beans and other produce ready for customers early on a Saturday at the Public Market. Cheryl has worked at the market since she was a kid.

"I had friends here," he says.

He worked nine years at the former Duffy Motts plant in Holley, testing cider and other drinks. Then he worked 20 years at Kodak in statistical control. While at Kodak, he also attended the Colgate Rochester Divinity School for three years, earning his degree in ministry.

Moss pastored a church in Albion, an African Methodist Episcopal Zion Church, for eight years, before leading an AME Zion church in Jamestown for another eight years. He would drive more than two hours each way to serve that congregation. (He still serves in ministry as an elder at the Memorial AME Zion Church in Rochester.)

Moss also served as chaplain of the Orleans County Jail for about 30 years. With all his responsibilities, not to mention raising nine children with his wife Betty, he has also sold produce at the market. With Moss's Fresh Fruits and Vegetables, he grows the crops and networks with other farmers and wholesalers to ensure a variety of items each market day.

I tell him I don't know how he did it, or does it. He says it's been "a family affair," with lots of help from his wife and children.

"It's also been fun," he says after buying me an egg-and-cheese sandwich at the little diner at the market. "It generates funds, and you meet a lot of people."

Moss started selling at the market after some coaxing from the Kirby brothers in Albion. Moss was picking apples part-time for John, Bob and George Kirby. They were vendors at the market and they thought Moss would be a good addition to the marketplace — and he might be able to make himself some extra money. Moss's stalls at the market have been across from Kirbys' ever since.

Early rise

Moss didn't want to make a big deal out of his workload. He wouldn't say how early he gets up or all the effort involved to get his truck loaded with a selection of produce. "People really don't care about that," he says.

On Friday, a day before I would join him at the market, Moss tells me to get there early or I won't find a parking space nearby. He tells me he will be there around 5. It's ridiculously early by my standards, but I say I'll be there at 5:30ish. He says that's fine.

I set a personal record Saturday in getting up at 4. I'm at the market in pitch darkness at about 5:30. Moss and his daughter Cheryl already have the tables out and are beginning to fill trays with several varieties of beans, peppers and onions, as well as other things, some I have never seen before, such as okra and some kind of turnip.

I get a rundown of the prices, which range from a buck to $1.50 per tray. Some 8-pound onion bags are $3 each and others are $4. I'm not sure which varieties are which. I just know the ones on the right side of the table are 3 bucks and the ones on the left are $4. There are 8-quart baskets of beans and peppers that vary from $6 to $7.

The Rev. Moss peels the outer layer off of onions to give them a fresh look for customers at the Public Market.

We are nearly set up when the wave of customers hits around 6 in the morning. It's a rush of humanity like I've never witnessed before. I can only compare it to a busy airport, except at the market there aren't wide lanes to walk down. People seemed jammed together.

I have most of the prices straight, but sometimes I have to ask Cheryl. Some customers also wonder if I have recipes for the beets and the very popular cranberry beans, a pinto bean that would be the top seller for the day. I ask Cheryl to step in with recipe suggestions because I'm clueless.

Cheryl, a labor and delivery nurse, has plenty of advice for customers. She says she likes the market and working with her parents and sisters. "This was our part-time job for school clothes, structure and discipline," she tells me.

There isn't much time to chat with the Moss sisters. It's a steady rush of customers.

Ruth, Cheryl's sister, works as a school teacher in Rochester, a second career after working as an accountant. She likes learning recipes from customers and getting a taste of their cultures.

Cheryl Moss, left, and her sister Ruth restock the containers of peppers, carrots and other vegetables in a business, Moss's Fresh Fruits and Vegetables, they run with their father.

"You see people from week to week and year to year," Ruth says. "That's kind of cool. You meet so many different people."

A capable market man

I think I'm doing a decent job. I'm friendly and polite with the customers, although sometimes I'm too chatty. One lady has a Rochester Women's Boxing League T-shirt on. I suggest she get a big helping of jalapeño peppers to help with her training. I'm the only one to laugh.

I'm keeping the produce replenished in the trays and baskets. I look for any open spots and refill them. Rev. Moss has a few requests for me. He usually wants more bags of onions to prep for sale. He peels off the outer layer, so the onion looks fresher and more vibrant.

He doesn't bother too much with tomatoes and beets. He thinks customers prefer seeing a little dirt on them. "That way they know they're fresh from the farm," he says.

My biggest struggle is just getting the plastic grocery bags open. They seem to stick together. If there is a break in the action, I get some bags open so I don't have to fumble with them when there are customers waiting.

Rev. Moss eases off on the pointers after a couple of hours. He gives occasional frowns when I'm standing with my thumbs in my back pockets.

I'm not resting, I tell him. I'm just scanning the crowd for potential customers, thinking a little eye contact may draw them in.

"You'll scare them away," he says. "They know what they want."

Why stop?

After five hours of selling, the Rev. Moss insists I wander the market and meet the other local folks at the market. I see about 10 other farms or flower shops there from Genesee or Orleans counties. They all love the bustle and brisk business at the market.

The crowd seems to taper a bit by noon, and at 2 p.m., Moss tells me to go home. He urges me to get some peaches and produce from some of the vendors. I would buy bags of plums, nectarines and peaches.

I again tell Rev. Moss that he sets an inspirational example with his hard work.

He sees no reason to live differently, even though he'll be 80 in three years: "As long as you're physically able, who says you have to stop?"

Chapter 13

Picking plump pumpkins is no treat

Tom Rivers joins a crew from Root Brothers Farms in harvesting pumpkins from a 25-acre field in late September. The pumpkins were huge from the plentiful rain and hot weather in 2008. (Photos by Rocco Laurienzo)

BARRE — It looked like Root Brothers workers were just throwing a football around when I stopped by a pumpkin patch in Albion in the fall of 2007 for a story about harvest help for local farms.

There were lines of workers flinging 10-pound pumpkins down a row. The pumpkin processional reached workers on a wagon, who placed the pumpkins in sturdy and tall cardboard bins. It looked like a lot of fun.

I made a mental note a few months ago to call Robin Root, co-owner of the farm, in late September, asking for a chance in the pumpkin patch. I figured it would be a welcome change of pace from the grueling days with cabbage and cucumbers.

Robin said I could give it a shot. He said the workers would treat me well as long as I didn't have any bad tosses or drop any pumpkins. Basically, don't do anything to slow down the work, and everyone will be happy.

On a warm Monday about a month before Halloween, I got my chance. I was riding the high from the Buffalo Bills' thrilling come-from-behind win on Sunday. I was ready to toss the pumpkins, figuring I could do a decent impersonation of Bills QB Trent Edwards.

But these pumpkins, fueled by so much rain this year, were enormous. Forget the happy-go-lucky toss with these big orange blobs. Just getting them off the ground and hand-delivered to a worker on a wagon took a mighty effort. I'd guess some of these pumpkins topped 75 pounds, maybe even 100.

Long clippers

I arrive at the pumpkin patch around 10 in the morning on a Monday in late September. Robin had told me by phone Sunday that the workers would be trimming cabbage early before moving to the pumpkins. When I get to the pumpkin field at the corner of Wilkins Road and Route 31A, there are about a dozen workers from Mexico out in the field. They have clippers with 3-foot-long handles to cut the stems on the pumpkins.

There's a glitch. Robin forgot to tell Benito, the leader of the crew, that I was coming. Benito says no problem. "We take anybody who wants to work."

He hands me some clippers and tells me to cut the stems so they are about 3 inches long. It seems like an easy job with these long clippers. We don't even have to bend over. I expected we'd be on our knees with hand clippers. The long cutters are a new addition this year at Root Brothers. Robin got a deal at Wal-Mart, an end-of-season special. They were only $2 more than the hand clippers.

"The guys don't have to stoop," Robin tells me a couple days later. "It makes it so much easier on them. Why create more work for them?"

I figured the stems would be tough to cut through, but they don't put up much of a fight. They even seem a little soft. Benito says the pumpkins will be left out until the next day so the stems can harden and dry.

Benito, 40, has worked on American farms since he was 17. He has been with Root Brothers the past four years. He doesn't like being called a crew leader or farm manager. "I like working with the guys," he tells me. "We're all treated the same."

Benito started his American farming career in Virginia, picking tomatoes and cucumbers. He has worked 12 years for Western New York farms, where he said the farms are always looking for help, and the pay is good, especially for the real tough jobs like cutting cabbage. A speedy cabbage cutter can make $200 a day and easily top $1,000 a week, Benito says. "But it's a hard job," he says. "Pumpkins are easier."

Benito still makes regular visits home to Mexico. He prefers life in the

United States. "In Mexico it's so poor," he says.

A new friend

Just like in other farm jobs this past year, one of the foreign workers almost immediately befriends me. A 23-year-old named Roberto comes up to me and wants to know my name. I tell him I'm Tomás, and he wonders where I live, if I'm married, and if I have kids. I give him a quick biographical sketch. I think he understands most of the story. (His English is a little shaky.)

Roberto tells me he's married, learning English, and this is his first year with Root Brothers. He spent last fall picking apples and he says he didn't like the job. He prefers cutting cabbage to climbing ladders and hauling 40-pound baskets of apples.

Roberto says the strap holding the heavy bins of apples would dig into his shoulders. He grimaces telling me about it. I tell him I'm going to give apples a shot some weekend real soon. "Pumpkins are easier," he says.

We keep clipping the stems, and at one point, Roberto warns me he sees a snake near my feet. I look down, but I don't see it amidst the thick vines and leaves. Roberto looks, and he spots it again. I move to a different spot.

Workers for Root Brothers use long clippers to snip the stems on pumpkins.

We spend about two hours cutting stems, and I don't think I've broken a sweat. The hardest part is walking through the fields. The pumpkin vines sprawl all over the place. There aren't obvious rows to walk through. I nearly trip several times from the vines. Around noon, we break for lunch.

A feast with some spice

A worker named Jesús has a good fire going under the lone maple tree near the corner of the two roads. Jesús, 19, surrendered his clippers at the beginning of the shift so I could try the job. He spent the next two hours building the fire and helping to fold mammoth cardboard boxes, which are double the strength of thick cabbage containers.

Benito, left, cooks pork with some of the Root Brothers workers. Benito, 40, has worked on American farms since he was 17.

Benito tells me today is a special lunch. The workers on Friday paid about $200 to have a pig slaughtered. On Saturday they prepared the pork to their liking. There's a metal tin that looks like a lasagna pan loaded with pork scraps. The workers throw them on a grill placed over the fire.

They also warm up some tortillas. Roberto and Benito insist I try one. I grab a slender piece of pork and a heated tortilla. Benito urges me to try some onions, a pepper and a little of some sauce the guys whipped up.

It's all new food to me, and I want to be polite, so I give it a shot. I take one bite and everyone waits for my reaction. The hot pepper and the sauce are very spicy, and I have to sprint to my car for water.

It's not too bad though, as long as I get two gulps of water after every bite. My main complaint is the pork is hard to chew. I ask Benito if the meat should be so hard and he looks at my selection. He says I picked a mostly skin piece of pork. He tells me not to eat it and try another piece.

Another worker named Rene is watching this and he points to a long, thick piece of pork. It looks like the best offering in the whole lot. Rene insists I take it. I'm a little reluctant to take the choice piece, but I do, and quickly I consume it. It's good, especially without the hot sauce.

Lunch is almost over and I see one worker fidgeting with a brace for his back. I don't take that as a good sign. Apparently we're in for some serious work.

Plump pumpkins

Benito announces that one group of workers will be going to the far corner of the 25-acre field, cleaning pumpkins and setting them in rows. The other group will be loading pumpkins into a wagon at another part of the field.

I figure the loading will be all fun and games, a chance to throw the lightweight pumpkins around. I want to be with that group. We head about 50 yards away, high-stepping the crazy vines. I'm with four other guys, including Jesús, the fire maker.

He takes a deep breath. He speaks a little English and he says we'll only be going after the heavy pumpkins. Root Brothers wants the big ones out of the field.

Jesús climbs aboard the wagon, which is pulled by a tractor. Rene is driving. There are two rows of four tall cardboard bins on the wagon. A guy named Ignacion, who looks to be in his 40s, and I start grabbing the big pumpkins. Ignacion tells me to leave some if they are too green or if their stems are too short.

Tom Rivers nearly pops a blood vessel in his neck trying to lift one of the heavy pumpkins at an Orleans County farm.

I am stunned by the weight of these pumpkins. The 40- to 50-pounders aren't too bad, but I have to do a deep squat and spread my arms wide to grab some of them.

A few seem to weigh 75 pounds. There's a Mexican guy on the other side of the wagon and he is staggering with some of the pumpkins. This guy is about 5 feet tall and weighs 100 pounds. He needs help from the guy with the back brace just to lift the pumpkins over the top of the tall bins.

Jesús tells me I'm strong when I hand him one particular whopper of a pumpkin. Even Jesús, who's 19 and seems to be the toughest guy out here, grunts carrying this one.

Jesús wants us to throw some of the pumpkins to him, to pick up the pace. If they're 25 to 40 pounds, I can throw them a few feet. Jesús puts his hands out, showing where he wants the pumpkin, and I send the pumpkins flying. It's not exactly tossing the football around in the backyard, but the pumpkins-in-flight add some excitement to the chore. We get the bins filled and I take a turn on the wagon.

Pumpkin stacking

Jesús and Ignacion grab the heavy ones from the field and hand them to me. I set the pumpkins with their stems up in the bin for the first row. Ignacion demonstrates the second, third and fourth rows need to have the pumpkins upside down. Don't set the pumpkins directly on top of each other or the stems will break. I get the hang of it pretty quickly, although it can be tricky to cram these mammoth gourds in some of the corner crevices.

Jesús and Ignacion seem reluctant to throw the pumpkins to me but I encourage them to try. I can handle most of them. But Jesús chucks one when I'm not quite ready and it makes a direct hit in my ribs. I hear something crack and I'm convinced I just broke a rib. It's actually a snapped stem, but I feel a little wheezy from the blow. I keep on going.

We trade off about every 15 to 20 minutes, the time it takes to fill a wagon. Sometimes we help the two guys on the other side if we get our four bins full before theirs.

There's one particularly huge pumpkin that gets my attention in the field, and I'm determined to be a show-off and try to lift it. This is a pumpkin on steroids, and it takes everything I've got to get it off the ground. I'm convinced Jimmy Hoffa's body is inside. I stumble over to Jesús, and it takes both of us to get this thing into the bin.

Jesús watches Ignacion carry a pumpkin above his head at Root Brothers.

We probably fill eight wagons before Rene and a second driver named Andres declare that's the end of the giant pumpkins.

Robin would tell me later the pumpkins would go to Tampa, New York City and other markets out West. He would struggle to find buyers for them because the pumpkins are so big. Wal-Mart and the chain stores prefer 10- to 20-pounders, Robin says.

"These ones are hard to handle," he says. "They're twice as big as they need to be. We'll have to go to the farm markets where the kids like them."

Robin again insists the pumpkins top out at 40 pounds, but I just don't believe it. These things had anchors inside them.

Fetch and clean

When Rene takes the last wagon away the five of us pumpkin loaders head for the back corner of the field, to help the other workers clean and arrange pumpkins in rows.

I'm relieved the pumpkins are smaller, maybe 20 to 30 pounds.

Jesús hands off a pumpkin to Tom Rivers. Root Brothers Farm grabbed the heavy pumpkins, 40 to 50 pounds and heavier, and packed them for farm markets because they were too big for grocery stores.

They're all over the place, and Benito says we need to make orderly rows for the tractor and wagon to pass through. Getting the pumpkins lined up and cleaned will pay off with a speedier load later, he says.

Benito gives me an old rag shirt and fish net cloth. I'll use them to wipe mud off the pumpkins. He says orange and green pumpkins are both fair game. There are at least a dozen of us workers and we're scattered in about five rows, with people fetching pumpkins and setting them in rows. It's after 3 p.m. and some of the guys look near exhaustion. The tired workers stay in the rows and wipe the pumpkins clean while other workers fetch them from the field.

I'm riding a rare mid-afternoon high and go about the work with zest. I start my own row. I'm retrieving and wiping pumpkins, even whistling a little bit. A few pumpkins don't have their stems clipped, so I just use my car keys to weaken the stem. I just jab at it a few times and then break it off. It's far from a clean cut, but I have to improvise.

I notice some zigs and zags in my row. I can't seem to get a nice crisp line.

Roberto comes over and smiles at my attempt. "Your row looks like a snake," he tells me. "It's OK. We'll fix it."

I wish I had GPS technology for my sneakers, to help keep my pumpkins in a straight line. I just accept there are some things I'm not good at. I continue to fetch pumpkins with gusto, bringing them to Roberto, who gives them a final wipe and sets them in a better-looking row.

Roberto asks if I'm tired, and I tell him I'm doing just fine. He declares pumpkin work is fun, but I say I don't agree. It's not as bad as cutting cabbage and picking cucumbers, but certainly not fun, I tell him.

At 5 o'clock we get a half-hour break. Rene gives a few of us a ride back on the tractor to the main spot under that maple tree. My car is there and I have a few nutty granola bars in my backpack.

I drive back down 200 yards to where Roberto, Jesús and two other guys are sitting next to the row of trees. I have two granola bars left and I give one to Roberto and one to Jesús. They break them in half and share with the other two guys.

Jesús starts calling me "amigo" and wonders if I'll be back tomorrow. I say I won't, but I will finish the shift today, which ends at 7.

Seeking a straight line

By 5:30 we're back in action, although three or four workers are snoozing in the field. Jesús jostles them and soon we're going at full strength. Rene is back with the wagon and a group of eight workers are flinging and loading pumpkins, about 20-pounders, into the big bins.

It seems no one wants to fetch and wipe pumpkins so I take that job. Roberto isn't there to help me keep my row straight and it becomes clear I just don't have the knack for getting these things set in a straight line. My row seems to fade out and weave back inside. I don't know what my problem is. Two other guys come over to wipe the pumpkins while I fetch more. I notice they also try to reconfigure the row.

By 6, the sun casts a hard light on us. It will be sundown within an hour. Many of the workers seem to have lost their spunk, especially the five of us who are still fetching and wiping pumpkins. I know the end is near, so I muster a late surge of activity. My back is sore, but I want a strong finish.

The pumpkins need to be lined up in rows so a tractor can get through the field.

I start a fresh row and this time I'm determined — and hopeful — I can get the pumpkins in a straight row. I get 10 in a row, then 20, and things look good. I'm all by myself, retrieving and wiping the pumpkins and I keep at it for 45 minutes before I hear someone yell, "Tomás, it's time to go home."

It's Jesús, who's on the wagon with most of the other workers. It's getting dark and I check my row of about 100 pumpkins. They're in a perfect line.

Chapter 14

Eating humble pie in the orchard

Jesús "Chuy" Vallejo, left, inspects some of the Ida Red apples picked in early October by three workers, including Poncho, right. The apples, nearly perfect, will be used in Watt Farms gift boxes sold during the Christmas season. (Photos by Tom Rivers)

ALBION — The farm manager calls it "a special job," and many of the other apple pickers don't seem happy that I get asked to do it.

They'd rather I join them crawling on the ground, scooping up "drops" — apples that have fallen off the trees at Watt Farms in Albion. The drops seem to be everywhere under the trees and it looks like a daunting job to have to pick them up and place them in 20-bushel bins. They will be turned into apple juice or sauce. The workers are assured I will get my chance at the drops.

But for most of a Saturday morning in early October, I will be joining Poncho and Roberto, two of the 14 apple pickers at Watt's, in finding the very best Ida Red apples in the orchard. It's a slow process, and Jesús "Chuy" Vallejo, 35, tells me to only get the apples that are deep red, without a blemish. These apples will go in Watt's gift boxes from late October until Christmas.

Watt's sells its best fruit in gift packages that it ships throughout the country. Some customers will pay nearly $1 each for the perfect apples to be given away as presents. These apples fetch top dollar and Chuy — pronounced like "chewy" — tells me I need to do a very good job.

He gives me a bag that holds up to a bushel of apples, or about 40 pounds. The straps for the bag wrap around my shoulders and the bag hangs down by my stomach. I've been warned in previous months by other farmworkers that the straps will dig into my skin. The weight of the apples also will strain my back.

Even Chuy, who I met in late June picking cherries at Watt's, told me apples would be far more difficult than cherries because of their weight. I geared up for this job for more than three months, trying to turn myself into a lean, tough, apple-picking machine. I've been running and even hitting the weight room.

A little twist

Chuy first shows me the trick of picking the apples without ripping off the buds that are critical for next year's apple crop. He also doesn't want the stems removed or the branches snapped off during the picking process.

Luckily, those of us with the "special" task of getting the gift apples don't have to pick with furious passion. It's tricky to go fast and not hurt the tree and bruise the apples.

I'm a little slow in learning the picking trick, however. Chuy says I'm getting too much of my palm on the apples. I only need to hold them nestled between my thumb and index finger. Just apply a little pressure, lift up a bit and give it a twist.

Chuy can easily do this with his left or right hand. I continue to grip too hard and Chuy grabs one of my Ida Reds and holds it up in the sun.

"Do you see the bruise?" he asks me.

Not really," I say.

He shows the slight dent to me, which I caused by squeezing the apple too hard. He says it will become a brown spot in a day. He tosses the apple on the ground and tells me to try again.

It takes me a good 20 minutes before I get the technique down, lightly holding the apples and giving them the little twist, without destroying the buds and branches. Chuy seems satisfied with my work and leaves me with Poncho and Roberto.

Roberto looks to be in his early 20s and Poncho must be around 50. Roberto gets the task of picking the apples high in the tree. He needs a 16-foot-high ladder to get the high ones. Poncho and I patrol from the ground.

Some of the apples have dark scabs where they were pelted with hail. Watt's is one of the many local farms that saw much of its fresh fruit damaged by hail storms this summer. Watt's was hit by at least three storms and Chris Watt, co-owner of the farm, says the storms wiped out two-thirds of his orchards. The hail-scabbed apples are only good for juice or sauce, which commands less than a third of the fresh market price. But there are still enough good apples for the gift boxes.

Some of the near-perfect Ida Red apples await to be picked.

'Not good'

I confess the 40-pound bag takes a toll and I set the thing on the ground after a couple of hours. I try a different tactic, filling my sweater with about 10 apples at a time. Then I carefully set them in the bag. That way I'm not lugging around 40 pounds.

Poncho observes this technique and scolds me. "Don't let Chuy see you do that," he says. "Not good." I put the bag back on and try to toughen up.

Poncho is nice and friendly, and tries to communicate with me despite my limited Spanish and his limited English. At one point he reaches for a fresh 20-ounce Pepsi. He opens the bottle and offers me the first drink. I'm shocked by his gesture, but I decline the offer. I typically don't give my children the first drink from my soda. Poncho offers his to a stranger.

Poncho, Roberto and I fill two of the 20-bushel bins with perfect Ida Reds. We do it in about three hours. Poncho drives the tractor, taking the 800-pound bins of apples back to the cooler at Watt's.

New task

It's lunch time and I head to Watt's Farm Market where Karen Watt promised me a free lunch. Watt's is serving barbecue ribs, and I get the first plate. I also have some pumpkin spice coffee.

After lunch, Chuy leads me to the other workers who have been toiling with the drops. Chuy says it's an awful job. The workers seem to be in groups of one to four people, spaced in different sections of the orchard.

We drive past a woman who started just a week before. Chuy looks down the row, and he's happy to see small piles of apples under some trees. The lady isn't grabbing every apple and throwing them in the 20-bushel wooden crate.

Chuy says the workers need to pass on rotten apples. But some workers go so fast and grab everything because they are intent on boosting their pay with a Herculean output.

"See those apples," he says about the ones the woman has left behind. "She's doing a good job. When you see the ground clean, that's when you have problems."

Orchard queen

Chuy takes me to his wife, Esperanza, who is overseeing the drop operation. Esperanza is a striking woman wearing a pink baseball hat and pink sweater, with jeans and some dirty boots. She shakes my hand and shows me to my two rows of McIntosh apple trees. Chuy tells me to leave the rotten apples alone. The others can be snagged for processing.

The job is a real grinder, perhaps even worse than picking cucumbers. I quickly fill the 40-pound bag and it seems to weigh a ton. I lug it to the bin 10 yards away. It's nice not to worry about bruising the apples. Instead of gently letting the apples into the bin, taking care to avoid hard bumps among the fruit, I just dump the apples — destined for processing — into the crate.

Then I'm back to squatting under a tree, trying to avoid getting my eyes poked by branches. Or I'm on my knees going after apples with blemishes — scars from hail, insect decay, limb rub, scratches and other imperfections.

I'm happy to see these drops still serve a useful purpose. I've been in orchards before in the fall and seen the many little red dots of apples under the tree canopies. It always struck me as a waste. I didn't realize most of those apples would be scooped up for juice.

Poncho drives a tractor hauling a bin of Ida Red apples back to the cooler at Watt Farms.

This drop job isn't for those with weak stomachs. I grab many rotten apples and they ooze apple slop in my hands. There are lots of

A Watt Farms worker snags apples off the ground at the fruit orchard in Albion. With processing prices high in late 2008, the apple "drops" were worth retrieving for pasteurized juice and applesauce.

slugs and mud on some of them. Slugs get stuck on my fingers and hands and they're hard to shake off.

It takes me almost two hours to get my first bin full — 800 pounds of imperfect apples. Esperanza brings a tractor with a lift down the row. She inspects the apples, and says I'm doing a good job. She takes the bin and returns with an empty one. This lady — rugged and glamorous — seems out of place on the orchard. Maybe it's her makeup and pink accessories that don't jive with the grueling drop job.

Chuy swings by and he offers me an apology. He tells me I'm supposed to be shaking the trees, getting the last of the apples on the grass. He demonstrates the technique and about 25 apples fall to the ground. I have to retrace my steps and go back to a dozen trees. I grab branches and shake them, and apples fall to the ground. A few conk me on the head.

A spectacle

About every half hour, I hear bells ringing, a whistle blowing and the muffled voice of Chris Watt. He is driving a train through the orchard. Watt's brings people out on the train, giving a tour of the orchard. They also visit a U-pick part of the orchard.

When the train, which runs on massive tires instead of a track, gets close, I can hear Chris tell the crowd about the many hail storms that have battered fruit throughout the state. He says the hail has left fruit growers "with monumental challenges."

The superstars

It's probably around 4 p.m. when two Mexican workers show up about two rows away from me. I think I'm going at a good pace until I see these guys in action. They are like high-powered vacuum cleaners. They don't squat or kneel down to get the apples. They walk, bent over, grabbing apples with both arms. It doesn't seem to be a sustainable position, walking like that with 40 pounds hanging from your neck.

I just watch these guys and within minutes they clean out the apples from underneath a tree. Then they're on to the next one. They appear to be grabbing anything that resembles an apple. I am stunned by their pace.

Watt's Orchard Express rolls through the apple orchards at Watt Farms in Orleans County. The riders will get off and pick some of the apples as part of an agri-tourism initiative.

Chuy would tell me later these guys would each fill 10 bins of apples. A good showing typically is 5 or 6 bins of drops in a day's work. But these guys are the most determined of the crew, Chuy says.

At 5:30 Esperanza takes my second bin of apples and tells me the workday is over. It takes me more than two hours to do a bin, compared to the fast Mexican workers who can do one in about 45 minutes to an hour.

I feel humbled and slow, and beaten up by the day's work. My back and hips have taken a pounding. Chuy shows up, and I gingerly get in the van for a drive back to the farm market.

Tough decisions

I see Chris and he tells me the processing apples will play a bigger role this year because of all the hail. He was offered 7.5 cents per pound by one processing company, and Chris had to get the apples delivered to the plant. He rejected that and will keep his apples in storage until a better offer comes later. Watt's has its own cold storage and can wait for a higher price.

The farm also needs more workers. Chuy thinks the farm can meet its needs with six more apple pickers. Chris wants eight or more.

If he can't get enough workers, he will abandon the drops and focus on the apples on the trees, which are worth at least three times as much as processing fruit.

Chapter 15

Jamming with the Jamaicans (or nearly dying trying)

Duncan, better known as "Stretch," has no trouble getting the high Empire apples at Orchard Dale Fruit Farm while Kereon, right, picks the apples in the middle of the tree. The two are part of a Jamaican crew that spends three months picking apples at the farm along Lake Ontario. (Photos by Tom Rivers)

POINT BREEZE — When he's home in Jamaica, Duncan builds houses and renovates rooms. When he comes to Orleans County every fall to pick apples, Duncan transforms into "Stretch," one of the tallest and hardest working pickers at Orchard Dale Fruit Farm.

"Stretch" is 54, 6 foot, 2 inches, and has super-long arms. He doesn't need a ladder to get the apples near the top of the dwarf trees at Orchard Dale. The other workers just give Stretch a look when they're near too-high apples, and he gets most of them.

Stretch was one of seven Jamaican apple pickers I worked with on a Sunday in early October. Four of the workers were older than 50. All of them were much faster and more skilled at the job than I was. All were gracious to me, offering friendly pointers on preserving my back and getting the best fruit.

Introductions

I meet Eric Brown, co-owner of the farm and Brown's Berry Patch, at about 9 a.m. He says I can stay as long as I want, which to me implies I can leave by lunch if the job is too hard.

I've been looking forward to this day, especially because the last Jamaican crew I worked with, in August, was so spirited while cutting cabbage in a major rainstorm. I marveled at their work ethic and cheerfulness.

Eric takes me in a van to an orchard maybe 2 miles from the Brown's Berry Patch farm market at the corner of routes 98 and 18. The crew is picking Cortland apples. The fruit is big, in shades of red to white. Eric tells me to get ones that are at least 50 percent red.

He introduces me to everybody and tells them to "Treat him like another worker." Eric also helps me put on the bushel bag. It holds about 40 pounds of apples. Eric crisscrosses the strap on the bag, and shows me how to wear it so the weight is distributed evenly on my back.

I'm ready for action and Eric hands me off to the crew leader, Sherlock Leys, 47. He has worked for Brown for 17 years. He spends six months at Brown's farm and rest of the year working part time for a railroad and at other farms in Jamaica.

The leader

Sherlock tells me he likes running the tractors at Brown's, inspecting the fruit and directing the crew. He first came to the United States to work on farms when he was 21. He once cut sugar cane for five months in Florida.

"There was lots of bending in the sun," Sherlock tells me. "It was the worst job."

He acknowledges picking apples — "carrying the sack" — is hard. "No work is easy," he says.

Sherlock seems pleased that I'm doing a newspaper story about the Jamaican crew. He wants his wife and two children, ages 21 and 25, to know what he does at the fruit farm. He has used the money from the job to buy a house in Jamaica and send his kids to school. "I just want to make them happy," he says about his family.

'Just the red ones'

Sherlock is no pushover when it comes to picking apples. He won't tolerate apples that aren't red enough, that are too small, that have missing stems, or those with even the slightest defect.

He regularly takes 20 apple samples from the workers, and cuts the

Sherlock Leys, left, and Eric Brown look over some of the freshly picked apples.

apples open, looking for bruises, insect rot and other blemishes. His demands for perfection make me nervous. I know if I go too fast I'll likely bang the apples together, causing bruises, or rip buds and branches off the tree. I make quality my focus and don't worry too much about speed.

I fill my bushel bag and walk to the trailer holding five bins, each with room for 20 bushels or about 850 pounds. Four of the bins are for fresh market apples, the ones that get top dollar in the stores. Another bin is for processing apples, the apples that aren't quite ripe enough or have missing stems or other defects.

I gently open the bottom of my bag and let the apples slowly make their way onto the heap of fruit. Sherlock tells me to take it easy and try to smooth the apples' bumpy ride. He even guides the apples out of the bottom of the bag with his hands.

He grabs a few of my apples and says they're not red enough. He sets them in with the processing apples. "Just grab the red ones, Tom," he says.

Taking a sample

The Jamaicans are friendly with each other, often singing and laughing on the job. One worker, Ferdinand, watches me closely and also tells me I'm getting apples that aren't quite ripe enough. He says the workers will be swinging back through in a few days to get the apples that aren't ready today.

It's a little tricky to tell the color because the apples give off a natural bloom, a coating I had mistakenly thought was a spray. Rub the bloom off and the apples are either much redder or lighter in color. Ferdinand, who surprises me when he tells me he is 60, just says to go for the ones that are clearly a darker shade of red.

I'm pleased with my picking skill. I've become proficient at not snapping off buds and branches. But I'm not perfect. Sometimes I don't twist the apple off quite right and the stem pops out. Without the stem the apple won't hold up in storage because decay will start on the inside of the fruit. I find the stemless apples make a good snack and I munch on a few out on the orchard.

On maybe my fifth load of apples Sherlock grabs 20 apples from the top of the bag. He puts them in a tray and slices them open. I go back to picking.

With the following load I ask how my apples looked. He says 19 out of 20 were good. Only one had a bruise. "Not bad," he tells me.

A little ladder duty

After two hours the crew moves on to a new orchard with rows and rows of Empire apples. Eric says these are the easiest to pick, especially when they are ripe.

I ride on the back of the trailer in one of the bins. I'm struck by the neatness of the Brown orchards — no wild branches sticking out into the rows. It feels like Augusta, the famed golf course that hosts the prestigious Masters tournament.

I'm also relieved Eric doesn't seem to have any tall trees, the ones that require 16-foot-high ladders. I only see dwarf trees that might peak at 10 feet high. A simple stepladder does the trick to get the high apples on those trees (when Stretch isn't nearby to get them).

I avoided the ladder with the Cortlands and I also stayed off one the previous day picking apples at Watt's. I have a fear of heights that I was forced to confront in late June picking cherries on the long ladders. I make a beeline for the ladder when we first get to the Empires. I only have to go up a few steps to get the highest apples.

Ferdinand is one of the older workers at Orchard Dale Fruit Farm.

One of the Jamaicans named Kereon, 32, tells me I'm too far from the tree and don't have the ladder set securely. I quickly wrap up the ladder duty, climb down, and Kereon, a welder back home in Jamaica, says he'll take it from there. He uses the ladder to pick from some of the other taller dwarf trees.

I'm just happy to have the ladder chore out of the way. I can say I did it without my legs shaking or heart pounding – at least not too much.

Low-lying apples

I think I'm getting the knack of this apple picking. The Empires truly are easy to pick. I can get them one-handed with either my left or right hand. I think I'm going at a brisk pace, and I hope some of the Jamaicans are taking note of my skills.

But no one seems too impressed. All of the workers seem so professional about picking. They laugh and sing, but they are very focused on the job.

The one advantage I seem to have over these guys is a remarkable set of nimble knees. I can easily squat down and get the low apples. A few of the guys even ask me to get some of the low ones for them. I'm happy to be useful.

I typically will set the apple bag on the ground while squatting down and getting the apples. This seems to work just fine. I get a break from the weight of the bag, and I'm within striking distance of the low-lying apples, some only a few inches off the ground.

I think I've found my niche, but in the early afternoon, hours after perfecting this technique, Stretch tells me it's not good to set the bag down. The apples might bump together and get small bruises. He shows me his style. He just spreads out his legs, and bends over with the apples hanging near his stomach, putting an incredible burden on his back. He quickly grabs 10 or 15 of the apples from the bottom of the tree.

I try his style, but it's a back-killer. There's no way I can bend over with a nearly full load of apples. I wait for Stretch to move on before resuming my crouching position.

Clifford doesn't look his 52 years of age. He is nimble at climbing the ladders and getting the high apples.

The orchard walk

I think I'm keeping up a decent pace. I've also taken to putting some of the stemless apples or those with scratches or imperfections in the pocket of my fleece sweater. I put those in the processing bin and keep the best ones in my bag for the fresh market. I think this is ingenious, saving Sherlock some sorting time.

My biggest struggle is the walk to the trailer of bins. Sometimes it's a good 30 yards, and we apple pickers, with a full bag of apples, look nine months pregnant waddling through the tall grass. The walk with the bushel of apples takes a toll on my hips, but I'm feeling fairly spry, even into late afternoon.

I find the workers all start opening up to me. There's one guy, Clifford, who I am certain is in his mid-30s. He tells me he's 52, and I just don't believe it. He pulls out his Jamaican driver's license and shows me. He was born in February 1956. Clifford also got his start with American farms by cutting sugar cane. He says that was a grueling experience and he'd rather pick apples.

I find out more about all the workers, but Stretch is my favorite because he is such a hard worker. He has a quiet dignity and refuses to take it easy. You'll never see him set the apple bag down, for example.

I'm next to Kereon, who's only 32, when Stretch seems to just blow by us on the other side of the tree. "Stretch is just amazing," Kereon tells me. "I have so much respect for him."

Stretch's story

I try to get Stretch's story in the afternoon. I get bits and pieces because I can't keep up with him. He tells me about his construction work in Jamaica. He enjoys working on houses, "but it's on and off. It's really not steady."

So he comes up and picks apples for three months, something he has done for 10 years. He also had a six-month stint cutting sugar cane in Florida. "We make some money so we can take care of our families," he says.

He has used the farm jobs to put three grown kids through school. His daughter, 27, will soon be a medical doctor.

I tell him apple picking is a tough job. "It's a tough job, but it's a skill job," he tells me. "You have to maintain your instincts at all times. You can't let your mind wander."

I think I know what he means because there was about one hour in the afternoon when I was just humming along, picking apples with few mistakes and doing so at a fast pace. I was in the zone, but then I lost it. I was distracted and started yanking off buds and pulling out stems.

Stretch says the job requires constant concentration and determination. It also requires a temporary separation from his wife and children.

"You miss your family," he says. "But I handle it. I've been at it a long time. It's a job I have to do."

Chapter 16

Hurrying (and failing) to keep pace with the Haitians

Jayne Bannister, 11, runs the tractor while two workers from Haiti unload 30-pound baskets of apples. Patty Harper, back left, works as the apple quality inspector at Excelsior Farms. (Photos by Tom Rivers)

POINT BREEZE — The driver looks a little disheveled, her long black hair poking out wildly from underneath her hoodie.

She is carrying a few notebooks, with some wrinkled papers hanging out. Jayne Bannister, a sixth-grader, is part of the apple harvesting crew. She runs the tractor at Excelsior Farms, a 35-acre apple operation about a mile from the shore of Lake Ontario.

When she's not driving the tractor, Jayne sits on the seat and studies her vocabulary words, practices her penmanship and memorizes Bible verses from Luke, chapters 4 through 6.

"We're not a normal farm," Jayne's father Roger tells me at about 8 in the morning on a Monday in early October. "Our tractor driver is an 11-year-old."

The farm would differ in many other ways from others I've visited this year. The workers are served lunch by the farmer, offered ice water and even aspirin. "It's hard work and I appreciate what they do," Roger says.

Root of relationship

I've known Roger and his family for 12 years. We go to the Albion Free Methodist Church together and I once had nursery duty with Jayne when she was a baby. She screamed for the whole hour. I still feel slightly traumatized by the experience.

But Jayne has become one of my favorite kids at church and in the local 4-H program. I've seldom been more nervous than watching her, at the age of 9, try to control a 1,200-pound steer in the show ring at the fair.

This year she is being home-schooled, and I am happy to see she will be in the orchard with us, running the tractor that pulls three bins on a sled. Each of the wooden bins holds 20 bushels of apples, or about 800 pounds.

Jayne proves to be a deft driver, making wide turns with the tractor and sled whenever it was time to move to a new row of apple trees.

A test

For the past decade, Roger has hired a crew of workers from Haiti to pick his apples. The workers are resident aliens or permanent residents staying in Florida. His group of eight, who appear to range in age from 25 to 50, come to his farm for about 10 weeks each fall. Roger says he marvels at their speed and skill in apple picking.

I want to see if I could at least come close to their picking rate. I've been trying to get in shape the past six months or so, and I figured my recent farm jobs would give me a chance for a good showing in the orchard. Today my output would be measured, so I would see how I fared against the Haitians.

'Take your time'

I meet some of the workers at 8 that morning. I tell them I'm a reporter and I'm writing about farm jobs this year, trying to give readers a sense of what's required to do these jobs well.

They seem a little apprehensive. At least three of the eight workers from Haiti say they'd prefer to not be in the paper. The workers pile into a van and Patty Harper, the apple inspector, leads them to the orchard. Jayne gives me a ride on her four-wheeler.

Today we'll be picking Empire apples, which Roger declares are some of the easiest to pick. They are ripe, and they easily come off the tree with just a slight twist. Roger and Patty tell us to take our time and not to bruise the apples. They don't want the apples dropped in the baskets, squeezed too hard, or thumped into the wooden bins.

Two apple pickers form Haiti briskly harvest Empire apples at Excelsior Farms. A crew of nine picked 62 bins or nearly 5 acres of Empires on Oct. 6, 2008, a record-setting day at the farm.

Those little bruises will turn into brown spots, which won't make the apple buyer happy. People in the grocery store will skip these apples. I also know from past fruit-picking experience to take care removing the apples from the trees. You can't just grab the apple and pull, or else the buds — even the whole branch — could snap off, exposing the tree to injury and disease. Removing the buds also means fewer apples for next year's crop.

Patty reminds everyone to "take your time." She checks the apples for bruises, insect holes, scars, deformities, color and size. She wants red apples that aren't too small. She thinks the crop of apples looks particularly good this year. Roger, unlike many of the Orleans County apple growers, didn't get hit with hail this summer.

Dwarf trees and short walks

Roger worked on his uncle Pete Nesbitt's fruit farm in Barre before starting his own orchard in 1985. Roger, 52, has a degree in pomology, the study of fruits, from Cornell University. He grows the shorter dwarf trees, which means the long 16-foot-high ladders aren't needed to pick apples. I'm thrilled to see that.

We only need 8-foot-high step ladders for some trees, but most of the trees peak at 8 feet high. The workers can get the high apples just by reaching up.

I'm also relieved that Roger's crew uses baskets that hold 3/4 of a bushel or about 30 pounds. At Watt's and Brown's, the two other apple farms where

I picked, the baskets held a full bushel or about 40 pounds. Ten pounds makes a big difference.

I'm also delighted to see that Jayne stays close by with the sled of bins. I never have to walk more than 10 yards to dump my apples. The day before, at Brown's, I found carrying the 40-pound load of apples to be the most strenuous part of the day. Sometimes I had a 20- to 30-yard walk, maybe more. That walk with the heavy load of apples was a back-strainer.

But at Roger's, with the lighter basket and shorter walk, my back stays strong.

Penny an apple

I'm going at a good clip, and I'm pleased to see I have the apple-picking technique down. I can even pick with both hands. The previous two days I was mostly using my right hand.

I'm confident I'm at least competitive with the Haitians. Patty, the apple inspector, even tells me after a couple of hours that I'm "doing very well." She gives each worker a slip of paper every time one of us unloads about 30 pounds of apples in the bin.

The other workers don't say much to me, and it seems they are all determined to keep a brisk pace. They even walk back and forth to bins with purpose.

One of the Haitians asks me near lunchtime if I'm feeling tired. I tell him I'm not, but my hips are a little sore. He says the job is tough on his legs. That's all we say, and this guy seems to pick up the pace even more after our short chat.

I keep trying to shoo away Jayne's dog, Elle, a striking charcoal-gray Weimaraner. The dog makes the rounds with the workers, looking for a little affection.

Elle, a Weimaraner puppy, likes to keep company with the apple-picking crew at Excelsior Farms.

I'm curious how many apples fit in the basket. I'd guess there are about 50. I count them, and find out it takes about 80 to fill up the basket. I share the total with Patty and she seems surprised it's so many. I count again with the next

After a worker unloads a 30-pound basket of apples, Patty Harper, left, gives him a ticket. The workers are paid 80 cents per basket. The fastest worker, with 253 tickets, would earn $202.40 for the day.

basket, and this time there's 77 apples.

Roger pays the guys 80 cents for every basket, so they're getting about a penny for each apple. He says they can easily surpass $100 in a day.

It takes about a half hour to fill the three 20-bushel bins. Shortly before they're full, Roger goes to the barn and retrieves a sled with three more bins. He then takes the full ones back to the barn. He uses two tractors for the chore and he switches with Jayne in the orchard. He goes back and forth, while Jayne stays in the orchard. She only has to move the tractor about every 15 minutes, driving the tractor forward maybe 10 yards to keep pace with the workers who are slowly advancing through the orchard, stripping each tree of its apples.

A free lunch

It's around noon when Roger pulls up in a van. He opens the back hatch, and he has subs and soda for everybody. The Haitians flip their baskets upside down and sit on them while eating lunch.

Roger started buying them subs or pizza for lunch about two years ago. He says the workers seem to appreciate it. It costs him $50 to $70 a day, but he doesn't mind. When he first started buying them lunch, Roger says the workers seemed puzzled and unsure why he was doing it. But they soon realized he was showing his appreciation for them.

"One of the workers told me, 'Boss man, you take care of us, we'll take care of you,'" Roger says.

Patty has worked with the crew for 10 years. She thinks the workers try even harder, especially in the afternoon, since Roger started bringing them lunch.

Hurting hips

After a half-hour lunch, we return to apple picking. We all seem invigorated, and we're all picking with a passion. Physically I'm doing just fine, except for my hips. By 2:30 they are sore and it's hard to make the walk to the sled with the 30-pound basket. The other workers don't seem slowed or uncomfortable.

I give up trying to be anywhere near as fast as they are, although I don't think I'm being too much of a slowpoke. Patty wonders if I'm quitting soon, and I tell her that's not on my agenda.

I want to get through the day and see how I compare to the other guys. It's only my hips that hurt, and they're fine as long as I don't move. Despite slowing down a bit, I still think I'm doing a decent job, showing due diligence to not hurt the trees and the buds.

It's 5 o'clock when Patty declares the work day over. Everyone counts their slips, to see how much money will be coming. While the counting is going on, I hear Roger declare it a record-setting day with 62 bins of Empire apples. The workers picked almost his entire 5-acre collection of Empires.

Final tallies

Patty expects some big numbers from the workers — she's right. The top Haitian, a worker named Jean, picked 253 baskets. The "worst" Haitian picked 198. Jean will be paid $202.40 for his efforts. The slowest picker, $158.60.

I count my slips of paper and I'm just shocked. I'm expecting maybe 120 to 130 slips. I've only got 89, which is good for $71.20.

Jean worked about eight hours or 480 minutes. He picked 253 3/4-bushels, or one basket of 80 apples in about every two minutes. He was snagging an apple about every 1.5 seconds, and that includes the walk back and forth to the sled of bins.

Roger Bannister, right, oversees the apple harvest at his 35-acre fruit farm near Lake Ontario. The team includes his daughter Jayne, center; apple inspector Patty Harper, left; and eight workers from Haiti.

My pace was about one apple every 5 to 6 seconds. Patty declares that a "bad" showing.

"I thought you were doing better than that," she tells me. "We'll have to throw you off the crew."

I think maybe some of the slips fell out of my pocket. But I was careful, and I typically doubled-checked throughout the day to make sure I wasn't dropping any. I tell Patty I was following her advice to be careful and not bruise the apples. She notes the other workers didn't bruise them, despite going at such a fast pace.

Jayne, the 11-year-old, chimes in that I didn't do too poorly considering it was my first time. I tell her I've actually been at it a few days.

I think realistically I could get to 100 baskets with a brisker walk back and forth to the bins and maybe a more determined picking pace throughout the day. But I doubt I could get past 120.

Roger says he picks apples at least one day every year with the workers, to see the conditions in the orchard and to understand what he is asking of his employees.

I asked how he did. He says he picked 101 baskets about a month ago. "But I was driving the tractor, too," he says.

Roger says I shouldn't feel too badly. The disparity in the output is more a reflection of the skill and determination of the Haitian workers. "They do an incredible job," he says. "I watch them with amazement."

Chapter 17

Testing farm-built endurance on the marathon trail

Tom Rivers takes a jog along the Brown Street canal bridge in Albion during another December snowstorm. He lost 40 pounds during 2008. (Photo by Nick Serrata)

I thought this saga of farm labor stories was over after the apple-picking with the Bannister family and their crew from Haiti. I found out I could spend six months whipping myself into shape, learning fruit-picking techniques, and still only work about one-third as fast as the best professional picker, and only about half as fast as the slowest guy.

I thought the data from that job – my 89 tickets compared to Jean's 253 – emphasized how good these guys are, how much faster they are than an "average" American. That seemed a good ending point.

But there's more to tell.

In August, after the absolutely brutal day in the muddy cabbage field, I was certain no job could be harder. I was chatting with a friend about a week later who had just completed a marathon. The 26.2-mile run is held up as the ultimate endurance test. My friend finished the course in less than 3 hours, 30 minutes. The next day he was a little sore, but he seemed in good spirits.

A few days after about 10 hours in the cabbage field, I could still barely pick up my 2-year-old daughter, who weighed 23 pounds. My wrists still felt mangled from heaving the 20-pound mega-ton cabbages. I compared my condition — aching bones and tendons — to the marathon runner, who seemed to have no trouble going about his usual business the day after the race.

When I suggested that the cabbage job was far more difficult than the marathon, the extreme runner doubted me. "If you showed up, with no experience in the job, and got through it, it can't be that bad," he replied.

At that point in mid-August, I was covering 3 to 4 miles at a time on my own recreational jogs. I figured that was enough. I hadn't run farther than 4 miles in a decade.

But after that conversation, I decided to stretch it a little farther, see what I could do. I live in Albion, where the Erie Canal cuts through the village. There is a nice crushed gravel trail along the historic waterway. It's perfect for runners who don't like traffic. I would usually run a loop that was about 1.5 to 2 miles in one direction and then come back home.

I decided to extend the loop, run to the next canal bridge, stretching the course from Butts Road to Keitel Road. Then I pushed it to Densmore Road. And then to Transit Road. Within two weeks I was running 7 to 8 miles. With the newfound iron will I discovered in the farm fields, I found I could take a little pain at mile 4, push through it, get in a groove and keep going.

I still figured a marathon was a pipe dream for me. But I started entertaining the possibility on Oct. 18, after running a 10-mile race in Rochester to benefit Hospice. I was going at an 8-minute-per-mile pace the first nine miles. I felt my strongest the last mile, running it in under 7 minutes. I finished with a 7:49-pace. A couple days later I was on the Disney Web site, registering for the Jan. 11 marathon.

It was an impulsive move. I was going to do the half marathon, but the registration for that event was full. The marathon was 97 percent full. I didn't think I should hem and haw for a few days and risk being shut out. Disney has the added attraction of being near my brother, who lives in Melbourne. I wanted to visit him.

I didn't think I was being foolish by signing up for the race. I had dropped 30 pounds since March. But even more important than being in shape, my farm work experiences had unearthed a motivation and mental toughness that had been absent for a long time. I believed I could fight through anything after cutting cabbage and picking cucumbers. The farmworkers showed me I had a reservoir of stamina I didn't think existed.

My running friends and family members questioned my sanity. Going from running 10 miles to 26.2 in about 10 weeks was pushing it. The running experts say you should have a good year of serious running before trying a marathon.

You need to gradually build up your endurance. You can't do it too quickly or you'll likely injure yourself, which I discovered in early November when attempting a 12-miler. I developed "runner's knee," or the more scary-sounding Patellofemoral Pain Syndrome, a softening of cartilage around the knee. Those running experts suggest backing off the training schedule, maybe running every other day.

So that's what I did. But I still had to get some serious mileage in to have a shot at a respectable marathon. Going every other day, I cut out the shorter runs, the 3- to 4-mile stuff. Almost everything was 5, 7, 10 miles or more. I also ran them hard, about as fast as I could muster. I thought that might make up for not having a bigger base of miles in preparation for the marathon.

The day off between runs helped, and after every run I iced my left knee with a bag of frozen green beans. I also wore a strap just below the knee for added support.

The knee still hurt, especially on the long runs, but I wasn't going to let it take me down. However, it did add a wild card element to the Jan. 11 showdown in Orlando.

For seven of the 10 weeks I tried to lengthen the long run by a mile or two each week. I typically did the big runs on Sunday afternoons, which spared me from watching the Buffalo Bills' collapse. (The team started 5-1 and finished 7-9, missing the playoffs, yet again.)

I ran mostly along the Erie Canal Towpath, often in cold, drizzly weather with the wind whipping in my face and, occasionally, ornery dogs nipping at my heels. With the long runs, 10 more pounds evaporated in the two months after the farm labor experience. That put me at 150 – my long-lost college weight.

For my last long run, in mid-December, I wanted to go back to where it all started. In late April, I had spent parts of two days planting onions with a crew from Triple G Farms, stooped over, pushing onion plants into the dark muck soil for a good nine hours or so. I barely survived.

The onion stint was what had first motivated me to get in better shape. From then on, I was laced up and out the door five times a week or more, until I got the "runner's knee" thing in November. In mid-December I asked Triple G co-owner Guy Smith to drive me out to the red barn by the field where I had worked the previous spring. I met him at Tim Hortons in Albion at 6:30 in the morning. Guy was having his morning coffee with some buddies.

They all seemed bemused and bewildered that I was volunteering to be abandoned in the muck land, forced to make it home on foot, in sub-freezing weather, with a thick layer of snow-slop coating the roadsides.

Guy measured the distance on his odometer and it was only 12 miles from Albion. I would need to add 6 more miles. So I took a turn in Barre at the Van Lie Shout farm, hit a side road to Route 31A and worked my way back to the village of Albion. It was a slow, steady slog.

But after 2 1/2 hours, I stumbled through the side door of my house and reached for the green beans. The knee felt OK. I hadn't lost my mind. Bring on the big race.

A battle to break 4 hours

ORLANDO, Fla. – At mile 4 of the longest race of my life, I hit the panic button. The clock already showed 45 minutes, putting me on pace to finish the marathon in 4 hours and 30 minutes.

I needed to speed up and do it fast.

I weaved to the outside of the huge pack of runners and left the asphalt course at Disney World. For me to move up, I needed to run in Florida's pathetic grass, mostly dirt with a little green poking up.

There were too many runners jam-packed on the pavement. I didn't see any open lanes. The race wasn't going to plan and I needed to red-line it for a few miles, trying to make up time and not use too much energy for the last critical and painful miles. I had set a goal to finish under 4 hours, and these next few miles needed to be fast, maybe too fast, just to give me a chance.

***

Before the race even started, I had some work to do. Call it putting on the war paint or armor. I knew I was in for a serious test, so I made sure I would have some help along the way.

It was just before 5:50 in the morning, and I was standing near Disney's Epcot with about 17,000 other runners.

I clenched my right fist around a coin from Great Britain, a 1900 penny my grandfather brought home from his service in World War II. Grandpa John lied about his age and joined the Navy at 15. He was a great influence on my life, taking me fishing, playing cards and attending my Little League games.

He died of cancer in 1991 at age 65. He is buried in Frostproof, a small town in Florida. I kept his coin close for my first marathon. I wanted to think of him often, to summon some strength and perspective. Yes, a marathon is a mighty task. But the long jog wasn't Normandy.

I remembered my grandfather's advice from our card-playing days: "Don't play all of your aces early." I had to try to restrain myself from

Tom Rivers wore bracelets honoring friends and family during the 26.2-mile race. He also carried a coin from Great Britain, a penny his grandfather picked up serving in World War II. (Photos by Cathy Rivers)

running like a wild man in the beginning of the race.

I also tugged on two bracelets, a pink one on my right wrist and a green one on my left. I wore the pink one from the American Cancer Society to honor the many friends and family I've lost to cancer, including my wife's 35-year-old cousin Karin Faulkner. She had died less than a month before the race. I thought of her and Dennis Bateman from Byron, Howard Harding and Dave Millis from Albion, my friend Matt Jones from college and a high school friend named Scott Cole. I thought of another friend, Gene Smith of Barre, who has fought cancer for many years, withstanding the ultimate marathon.

I wore the green bracelet — the Children's Organ Transplant Association — in honor of Jenna Stothard, an Albion native who died almost two years ago at age 22. She never lost her spunk despite battling kidney disease. She starred in Albion musicals, taking her dialysis machine with her to play practices. She was a fighter.

I also wore another motivational item from home: a T-shirt from Torrey Farms. This was in honor of all the farmworkers who had shown me incredible willpower in the last year while planting onions, milking cows, and harvesting fruits and vegetables. I had a phrase added to the back of the shirt: "Si Se Puede — Yes We Can!"

Just moments before the runners were released, the Disney folks unleashed flares and fireworks from a bridge. I was only about 20 yards from the launching spot and my heart pounded from the spectacle and the booming noise.

I reminded myself of the famous quote by Hall of Fame Buffalo Bills coach Marv Levy: "Where else would you rather be than right here, right now?"

***

I had been warned not to start out too fast because it was highly

Torrey Farms gave Tom Rivers a T-shirt from the farm to wear during the marathon. Rivers added the famous quote by Cesar Chavez: "Si se puede – Yes we can!"

unlikely I would be able to sustain such a pace. "You have to respect the distance," I was told several times by my father-in-law Sid Bolton, finisher of 11 marathons.

I intended to go out at an 8-minute-per-mile pace, but was talked out of it a day before the race by my good friend John Butler, who also is a runner. He told me I likely wouldn't get through the race if I went out too strong. "You'll be in excruciating pain," he warned.

***

At mile 4, I squeezed my grandfather's coin and took off on the grass. I ran a 7:30-mile and still had about 8 minutes to lop off to give myself a chance at 4 hours. That's a lot of miles to run well below a 9-minute-pace. I started to alternate between the pavement and grass, staying on the pavement for the water and Powerade that was handed out by many peppy volunteers.

My three-letter first name was prominent on my bib number, No. 3292, and many volunteers yelled "Good job, Tom" and other praises. I'm glad I didn't have a 10-letter first name that would have been hard to see.

I admired the runners, and I felt their determination. "For Sam" was on the back of one lady's shirt. One guy wrote "4 Jan" in ink on the back of his calves. I saw many declarations like this on shirts, arms and legs.

There were runners from all 50 states and 50 other countries. I saw people with shirts noting the runners' allegiance to Brazil and Costa Rica. There was a group of runners taking turns carrying an American flag.

Most of these people were lean machines. I couldn't help but think of how far I'd come in only a few months. In June I couldn't get through the 3.1-mile race in Albion without stopping a few times. On Oct. 18, I passed the 10-mile barrier for the first time. I got as far as 18 miles about a month ago.

About a week before this race, *The Daily News* ran a preview of my attempt at my first marathon. I told everyone I was determined to break 4 hours. We solicited guesses from the public about my time. Whoever guessed closest won a chance to go jogging with me. I expected maybe three people to enter this "contest." But 16 people sent in guesses, and nearly everyone predicted a time over 4 hours, with one over 5 hours.

Even though I had fallen behind the 4-hour pace, I was determined to surprise some people, those who predicted a slow time. I felt strong strong and I believed I had the willpower to endure. And don't forget I had the redheaded feistiness going for me.

***

I saw another passing opportunity at about mile 6. There was a long incline, gradually leading to an overpass. I ran up the hills leading to canal bridges for months, making me well-trained for the incline. I sprinted up it as fast I could, whizzing by at least 50 people.

"That's from the people of Western New York," I muttered under my breath. I don't like Florida for luring away too many friends and family from WNY. They fled for Florida's jobs, great weather and low taxes. I looked to sky, hoping a snowstorm would break out at Disney during the race.

I settled into a nice 8- to 8:30-minute pace over the next 10 miles or so. By mile 10, I knocked my overall time below the 9-minute pace, on mark to break 4 hours. But I wanted more 8:30-miles because I knew it was going to be slow-going at the end of the race. I needed to build a cushion so I didn't have to run fast at the end.

We ran on Disney's perimeter roads for about 10 miles, even passing by the Magic Kingdom's sewer plant, before we started seeing some famed Disney attractions. We encountered Donald Duck, Mickey Mouse and many Disney princesses (I recognized Snow White but don't ask me about the others). I high-fived Donald, some girl Duck, and many in the long lines of cheering spectators. Some runners carried cameras, and they stopped to get their photos taken with the characters.

I stayed on the outside edge of the course near the people. Many yelled my name (it was so easy to see on my bib) and extended their hands. It was a great thrill, even better than running through the castle at Disney.

As the race moved past the midpoint, there were more spectator spots, and it made a difference. The people were loud and many held encouraging signs: "You're awesome," "Yes you can," and "You inspire me."

There were four or five high school bands along the course, and they played heart-pounding tunes, including the Rocky theme song. I gave them a thumbs-up and kept moving.

***

I didn't start slowing down until maybe mile 16 and it was just a slight downshift. I was still under the 9-minute pace. I tried to hold off "the wall" as long as possible. I knew it was coming and I wanted it to be close to the finish line.

It happened around mile 20, the aching in the legs. I thought of the farmworkers, who just kept going, 13 hours cutting cabbage in the deep mud the day I joined them in August. I could only take 10 hours.

I also thought of a book I saw in the bookcase at my house. I could see the cover of the book, *Failure is Impossible*, a story about suffragist Susan B. Anthony. I spit on the ground and found some more firepower. I needed at least two more decent miles to stay under 4 hours.

I plugged along to mile 22, and I wanted the race to be over. I had about 8 minutes to play with over the 9-minute pace, so I could plod through the last four at maybe an 11-minute pace. That would be a slow jog.

I had been walking through the water breaks the second half of the marathon, and I extended that break from about 5 seconds to maybe 20. Then I tried to get my legs churning again. I thought of Bob and King Kong, the Jamaicans I worked with in the cabbage field. I thought of my kids and my wife Marsha, who was following this saga at home on-line.

After finishing the marathon at Disney, Tom Rivers mimics some of the Olympic athletes from 2008 by biting his medal.

One of my shoes had a chip and about every 5 miles Disney posted an update of the runners' times. I knew my wife could tell I was slowing down but was still within the 4 hour-goal.

I again recalled Marv Levy's quote, and I didn't want to regret how I handled these remaining miles. So I kept running.

People were withering, pulling off to the side to walk. I just kept pushing through.

It was around mile 24 when I told myself 4:02 would be OK. I started walking and accepted what would still be a respectable time. But then I saw Susan B. Anthony's book cover, and I again fired up the engine.

At mile 25, the runners were greeted by a gospel choir, singing boisterously, with flowing burgundy robes. I took one last short walk, knowing I could break 4 hours if I plowed through the last mile in about 12 minutes.

My legs felt dead, but I wasn't about to be denied. I gripped the coin again and started rumbling around a corner by amusement park rides.

There was a wild throng of people awaiting the runners. The crowd was screaming, waving and holding their signs. All the runners picked up the pace.

I just soaked it up, giving high fives to anyone who was willing. I got past mile 26, and I had more than 3 minutes to complete the last 385 yards. I passed the finish line in 3:58:51, giving me about a minute to spare.

I again thought of Coach Levy. There was nowhere else I'd rather be.

Comparing which is harder— Farm work or a marathon?

You may be wondering, "Which is more difficult?"

Now that I've done both — worked in the fields and run a marathon — I'm convinced cutting cabbage for a full day is more of a full-body physical smackdown. The cabbage job, and the cucumber picking one, will wear down your legs, your back, your arms — just about everything. And the day will never end, stretching 10 to 13 hours in the field.

With a marathon, I knew it could be over after 4 hours. When I finished the race at Disney World on Jan. 11, 2009, my legs definitely ached, especially my quads. I had a blister on my right foot that gave me a bloody sock, and two or three toes were particularly sore and had turned a light purple.

But the long run at Disney was easier because of the overwhelming support of a cheering crowd, the steady supply of water and Powerade, and the bananas and candy offered along the way. I was even given a medal.

With farm chores, the workers are mostly anonymous. There aren't cheering spectators heaping praise. No one is holding up a sign that says, "You're awesome!" like they do in a marathon.

The workers know the public is indifferent and sometimes hostile toward them. They've witnessed our country's heated immigration debate over the past three years. Ultimately, our country and our Congress find the issue too hot to touch.

That has left the workers with the added tension of immigration raids and other arrests by law enforcement. The workers try to keep a low profile. They're not looking for any acclaim, just a paycheck from their employer.

I ran the marathon with about 15,000 other people. Just before the race started, Disney launched fireworks and flares. They had high school marching bands along the course playing "Rocky" and other heart-pumping tunes.

We were given high fives by Disney characters and the enormous pack of volunteers. And we ran along a course carved through the famous park with a castle as the backdrop.

With the farm jobs, my boots were often caked in dirt or manure. I cut cabbage in a foot of mud, grabbed squash in a swarm of bees, picked cherries with centipedes crawling up my legs and milked cows while sloshing through liquid manure. I was thrilled to earn a little respect from my coworkers.

One way to measure the physical brutality of the marathon and farm work is the recovery process. With the marathon, I was still gingerly walking after two days and I tried to avoid stairs. But on the third day, I was in decent enough shape to play volleyball.

The cabbage job hurt my wrists and arms so badly it took me four days before I could lift my 2-year-old daughter. With the cucumber job, my whole body suffered for several days. My hips took the hardest hit.

With a marathon, runners are encouraged to take it easy for a week before even attempting a short jog. With farm work, you're right back out there the following morning.

You've got to be tough to run a marathon. You have to be tougher to be a farmworker.

Chapter 18

Seeing citizenship: Dream for many, fulfillment for few

Jesús "Chuy" Vallejo takes the oath of citizenship during a naturalization ceremony in Rochester on Feb. 19, 2009. (Photos by Tom Rivers)

I had thought for sure that my marathon chapter put the exclamation point on this book. My transformation into a long-distance runner led me to break a four-hour marathon. It was a daunting challenge, all that running, but picking cucumbers and cutting cabbage was far more difficult. It turns out, there's an even more climactic end to these tales.

I had hoped in my short stints with some of these farm jobs that I would make connections with the workers, relationships that might lead to other stories. My ultimate hope was to see one of the workers become a U.S. citizen. I thought that would be a good story.

The opportunity came in February. Jesús "Chuy" Vallejo had completed his citizenship requirements, everything but taking the oath. Karen and Chris Watt, owners of the farm where Vallejo works, said Vallejo wanted me to come to the ceremony and write about his two decades' journey from Mexican teen-ager to American citizen.

Here's the story.

The road to citizenship

ALBION — "You look like such a little kid," Karen Watt tells Jesús "Chuy" Vallejo. "You were so scrawny."

They are looking at the ID card Vallejo obtained about 20 years ago, soon after he first started working for Watt Farms in Albion. Vallejo, in the photo on his Green Card, is rail thin, with long, feathered hair. That was in 1988. Even he chuckles at the picture.

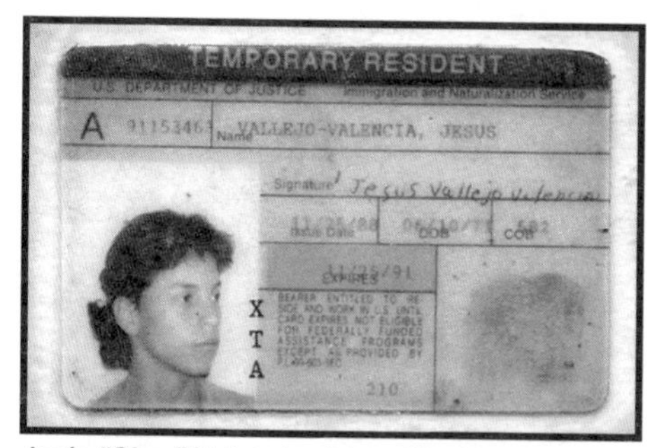

Jesús "Chuy" Vallejo is pictured in this temporary resident card issued Nov. 25, 1988.

"Now, I'm fat," he says, smiling and grabbing his belly.

Today, at age 36, Vallejo has gray in his temples. And the hair isn't as thick as it was when he was a teen.

A lot has happened in the 21 years since he arrived in Albion to work for Watt's fruit farm. Vallejo's uncle Leon was working for Watt's. He encouraged his nephew to travel from Mexico to Albion.

Vallejo arrived in the fall of 1987, his first time away from home. His first task in the orchard was cutting "suckers," sticks that stuck out of the bottom of trees. He helped bring in the apple harvest, then worked in the farm's cold storage, and trimmed trees in the winter and spring. He stayed about a year before returning home to Zamora, in the state of Michoacon, southwestern Mexico.

Vallejo kept coming back after visiting his parents and nine siblings. Initially, it was for the money. In Mexico he could work a full day and earn $2 — "That's nothing," he said.

In Albion he would marry — twice — and have four children. He also would become a valued full-time, year-round employee at the farm, moving up from a fruit picker to the orchard foreman.

"He's like a son," Karen Watt says.

And now, since a naturalization ceremony Feb. 19 in Rochester, he's an American citizen.

Working like a man

Vallejo is the oldest of five brothers and four sisters. In the Mexican culture, the oldest son has pressure to provide for the family. After second grade, Vallejo quit school and went to work at farms.

He joined his father picking strawberries, apples, potatoes, peppers and onions. When he was 14, Vallejo was making $2 a day for long hours in the fields.

Jesús "Chuy" Vallejo sits on a planter and sets a peach tree in the ground with help from two other workers in late April 2009. Vallejo works as a foreman at Watt Farms. He continues to take on more responsibilities at the farm.

"They paid me like I was a little kid, but I did the work a big man would do," Vallejo said at Watt Farms on Route 98 in Albion.

His father noticed the son's output and he argued for a pay hike for Vallejo. The bosses agreed, and they starting paying Vallejo a man's wages even though he wasn't even a teenager.

Vallejo's uncle Leon started working in Florida citrus groves before making the journey to Watt's to pick fruit. For about an hour's pay, Leon was doubling what his nephew made for a full day's work.

Vallejo was 14 when he made his first trip to Albion. He was illiterate and knew very little English. Leon had proven himself to be a good worker and a good man, Chris Watt said. He expected Leon's nephew would fit in, particularly with his uncle nearby.

Watt didn't realize Vallejo was only 14 when he first arrived in Albion. Vallejo, during a sweeping amnesty program in 1988, gained legal standing to be in the country.

The U.S. immigration laws haven't been significantly overhauled since then, and now an estimated 12 million undocumented workers are in the country. Recent efforts in Congress to reform the laws, allowing more legal workers for agriculture and other industries, have failed. President Barack Obama wants the 12 million people to be documented and "out of the shadows."

But with many pressing issues in the country — wars in Iraq and Afghanistan, and an economic meltdown — immigration reform isn't at the top of the agenda in Washington, D.C.

Mr. Watt said the current immigration system is "atrocious," often dividing families and creating a culture of fear for workers and their employers, who are vulnerable to immigration raids by police.

Even if a worker wanted to pursue the legal path in the current system, it could take years. "Once you get in that line, it never moves," Mrs. Watt said.

Ami Kadar, director of the Independent Farmworkers Center in Albion, said many farmworkers are in local counties with illegal papers. They have slim hopes of gaining legal status with the current immigration policies and attitudes in the country.

"It's really hard to qualify," said Kadar, who leads the group better known by its Spanish acronym of CITA. "If you're a farmworker and you've entered illegally and you have no family members here, I'd think it would be impossible."

Key employee

Vallejo spent his first Albion autumn picking fruit. In the winter, he stayed to help with cold storage and trimming trees. Tree pruning lasts several months, well into the spring. He helped plant trees in the spring, and by June and July, it was time to pick berries and cherries.

Vallejo stayed a year before he returned home for about a month. For two winters, he traveled with his uncle to pick oranges in Florida. Vallejo didn't like the job.

"It's too hot," he said about Florida. And the sacks of oranges "are too heavy."

Vallejo was happy to pick apples and trim trees in his teens. But by age 20, he wanted more responsibilities at the farm.

Chris Watt was willing to give Vallejo a chance. Watt made Vallejo a supervisor, responsible for managing workers, teaching them the picking jobs, monitoring the quality of the fruit and keeping the workers at a productive pace.

"He went from just doing things to wanting to know how things work,"

Jesús "Chuy" Vallejo, left, and a worker named Poncho trim apple trees as part of a spring pruning. Vallejo helps supervise the workers at the farm.

Mr. Watt said. Vallejo, by 20, also was speaking very good English. He said he learned mainly by watching TV.

Watt noticed Vallejo's potential and Watt recalled a conversation from about 16 years ago.

"We sat down with Chuy (pronounced like Chewy) and told him 'It's all up to you,'" Watt said.

He would teach Vallejo how to run the farm. Vallejo can do almost any job at the farm now. He said he knows every branch on about 40,000 trees spread over 250 acres at six orchards. Spraying chemicals to fight pests and disease is the one job he doesn't do. Vallejo said he doesn't like the odor. But Watt said he may need Vallejo to learn to mix the chemicals and run the spray rig.

"I'm getting older," said Watt, 61. "These body parts are getting worn out."

Watt was in the hospital for two days in late April 2009, when his back seized up. He normally runs the planter, and sets new trees in the ground. But with Watt's bad back, Vallejo was operating the machine, spacing new peach trees every 9 1/2 feet. The farm is trying to plant 8,000 new trees this spring.

Vallejo also is managing the apple tree trimming. He said the workers need frequent checks with the pruning, which he said is a difficult job requiring years to perfect. With branches jutting in wild directions, the limbs need to be trimmed to bring balance to the trees. Too many branches on one side and a tree will lean over. If there are too many branches on top of the tree, not enough light will pass to the lower branches, leaving the bottom layers with apples that never reach the right color.

Vallejo wants the bottom rows of apples to be the best ones. They're easier to pick for the workers. Many workers never gain the ability to quickly size up the tree, see it in three-dimensional form, and look for branches that need trimming, Watt said.

Holding a chainsaw for eight hours or more a day "wears out your arms," Vallejo said.

Vallejo's cell phone was going off every few minutes on Wednesday. He updated Watt on the progress with the trees. Vallejo answered questions from other workers and directed them. He also answered calls from his wife and his family.

Watt said he appreciates Vallejo's ambition and a sense of duty for his family. Since his first days at the farm, Vallejo has been sending big chunks of his paychecks to his family in Mexico. He wanted more responsibilities so he could earn more money, and send more home.

Path to citizenship

About five years ago Chris and Karen Watt told Vallejo he needed to be a little more selfish and stop sending so much of his earnings to Mexico. They encouraged him to buy a house.

Vallejo owned a trailer at Oak Orchard Estates. Three years ago he bought a house on Phipps Road.

Kadar of CITA said Vallejo isn't alone in sending money home. She often urges CITA members, which include about 1,000 farmworkers, to save more money for themselves.

"They live on very little when they're here," she said. "I tell them, 'What about you?'"

After Vallejo bought a house, Watt followed up that conversation with talk of American citizenship. If Vallejo became an American, he could better access loans and ensure a better future for himself and his family, Watt told him.

Jesús Vallejo poses with his family outside his home on Phipps Road in Albion. The group includes, from left: Jessy, 16; Tony, 9; Vallejo; his wife Esperanza; Victor, 11; and Marta, 13.

Vallejo was more interested in his family in Mexico. He knew if he became a citizen, he would have more pull with the U.S. government to bring them to the United States.

Vallejo said he also wanted to be a role model for his four children — Jessy, 16; Marta, 13; Victor, 11; and Tony, 9. The children are all active at Albion Central School, playing sports. Jessy, a 250-pound high school junior, plays on Albion's varsity football team, helping to anchor the offensive and defensive line. He wrestles at the 285-pound weight class and competes in shot put and discus in track. He said he wants to land a college athletic scholarship.

Marta, an eighth-grader, plays basketball, softball and volleyball. Victor and Tony both play youth football and baseball.

About a year ago Vallejo started taking weekly English classes at CITA in Albion. He honed his writing skills and studied American history, key elements to passing the citizenship exam.

Vallejo learned 100 facts of American history in sample multiple choice tests. When he took the exam in Buffalo last October, he was asked six questions. He passed, but missed one question. He was asked who is New York's governor. Vallejo answered Eliot Spitzer. But Spitzer had resigned by then and was replaced by David Paterson.

Jesús "Chuy" Vallejo holds his citizenship papers and an American flag while being congratulated by Karen and Chris Watt of Albion following a swearing-in ceremony Feb. 19 in Rochester. Vallejo has worked for the Watts for about two decades.

Vallejo said the process wasn't too difficult, but a bit drawn-out. He passed the test in October, but the Bureau of Citizenship and Immigration Services in Buffalo didn't schedule a swearing-in ceremony until Feb. 19 in Rochester.

That day, Vallejo joined 46 other people, including two others from Mexico, in taking the oath and pledging allegiance to their new adopted country. The 47 new citizens came from 26 different countries.

Alex Renzi, Monroe County Supreme Court justice, addressed the new citizens. Renzi noted his ancestors were immigrants from Italy.

"We are a stronger country because of you," he said at the naturalization ceremony at the Monroe County Office Building.

Karen and Chris Watt, and other employees at the farm, joined Vallejo for the swearing-in ceremony. He was given a small American flag as well as his citizenship documents.

Mrs. Watt said she was moved by the ceremony, which elicited tears from many of the new citizens.

"Every one of these 47 people has a story to tell," she said in the crowded courtroom. "The ability to dream is one of America's strengths and always will be its strength. It's the immigrants that see it that way."

Epilogue
'Paying a Price'

When I answered my home phone in October, after the series of articles on farm labor had been published, the caller thanked me for highlighting farmworkers and their contributions. The caller wanted the public to better appreciate the grueling work in the fields. But she worried I had made the workers out to be superhuman.

"They pay a price," Debbie Restivo, a public health nurse in Orleans County, told me.

Restivo has worked with farmworkers for 14 years, traveling to camps. She checks their blood pressure and screens for illnesses. She also gives immunizations and visits newborns. She often joins a team with the Oak Orchard Community Health Center's mobile clinic. They visit farmworkers at camps in Orleans, Genesee and western Monroe counties. Oak Orchard also serves the general population with offices in Albion, Brockport and Lyndonville.

The estimated 3,200 migrant farmworkers who spend 10 weeks or more in the three local counties often leave with muscle strains, fungal infections and many other injuries and illnesses, according to the Oak Orchard Community Health Center.

Restivo said many workers ask her for Motrin to help with their pain. Many also want vitamins for energy and creams to fight fungal infections on their feet. The long days, often working in wet socks, break down their skin and cause open wounds and infections, she said.

Restivo treated 661 farmworkers in Orleans County in 2007. The county in 2008 was approved for a $53,515 grant from the state Department of Health to provide services to the migrant population. The grant pays some of Restivo's salary, her travel expenses, bilingual literature for health tips, and services by Oak Orchard Community Health Center and the Genesee-Orleans Council on Alcoholism and Substance Abuse.

The workers pay $15 each for prescriptions of generic drugs, X-rays, exams and other care through Oak Orchard. Government programs, including Medicaid and the state grant, cover other costs. The workers are very good about paying their share, said Barbara Linhart, Oak Orchard's director of clinical support services.

Top issues

Oak Orchard in 2007 treated 1,328 farmworkers in the three counties for medical care, with upper respiratory, eye, ear and dental infections the most common ailment with 1,118 patients.

Gynecological care was second with 679 patients, who didn't necessarily have a medical problem and may have just received a Pap smear, mammogram or birth control, Linhart said.

Muscular-skeletal pain was the third most common medical issue, with many workers dealing with sore backs and knees. Oak Orchard will dispense pain killers, but they tell workers to be careful because the pills can make them light-headed and dizzy.

Ideally, said outreach coordinator Rosario Rangel, the workers would take a few days off and give their bodies a rest. But few do. "They're here to work," she said.

Many of the jobs require sharp knives and Rangel said workers may cut themselves or someone next to them. Oak Orchard treated 571 workers for injuries and trauma in 2007, with lacerations a frequent problem.

Rangel, whose husband cut cabbage for a Brockport farm for 20 years, said cabbage cutting, with its constant repetitive motions, is one of the leading causes of pain for farmworkers. They develop muscle and tendon strains, and sore rotator cuffs, as well as carpal tunnel pain in their hands. "It's so repetitive," she said.

Harvesting pumpkins, which often top 40 pounds, also leaves many workers with lower back pain, Rangel said. The most strenuous jobs, such as cutting cabbage, come with added monetary rewards, so workers want those jobs, said Ami Kadar, director of Independent Farmworker Center in Albion.

"It's hard working out there in the mud and the snow, but they can make more money," she said.

'Under the radar'

A non-profit agency, the New York Center for Agricultural Medicine and Health, calls agriculture "one of the most dangerous industries in the United States." From 1996 to 2000, an average of 34 people in New York each year died working on farms, with tractor-related deaths causing 43 percent of the fatalities. NYCAMH has been studying farm-related injuries the past three years, and suggesting ways to make the jobs safer, including using more ergonomic baskets that better distribute the 40-pound weight of the full apple bushel baskets.

The agency found in a study in 2005 that 68 percent of farmworkers "almost never" suffered injuries. But for the other 32 percent, strains and sprains are the most common problems, NYCAMH reported.

"Those injuries could end up as permanent disabilities if the workers don't get the treatment they need," said Jim Carrabba, an agricultural safety specialist with NYCAMH, which has an office in Cooperstown.

Temo Martinez and Katie DeFelice, outreach workers for Oak Orchard Community Health Center, stand outside a camp in Brockport with Oak Orchard's mobile health clinic. Oak Orchard partners with other health care agencies to serve farmworkers. (Photo by Tom Rivers)

He said few workers will take time off from work or complain about the pain.

"A lot of times they will keep working," he said. "They're afraid they could be fired. If they don't work, they're also not getting paid. They try to work as much as they can."

The farmworker population may not get even basic medical care because many workers "are reluctant to seek services" because they don't want to miss work or be sent back to their home country. "They're kind of under the radar," he said about farmworkers and their health care.

Oak Orchard is one of three community health centers in the state with services to migrant workers.

Difficult jobs

Many of the jobs, whether milking cows, picking fruit or packing vegetables, come with dangers and can take a toll on the body, Rangel said. "There's no easy farm job," she said. "They're all, in different ways, very hard to get done."

Despite the grueling nature of many of the jobs — 10 to 12 hours of bending over in the hot sun, rain, mud and sometimes snow — Rangel said workers can reduce injuries by being careful and being efficient with their movements.

Her husband cut cabbage until he was 62. Some of the fruit and vegetable workers are active on the farms even longer.

"We have farmworkers who are in their 70s out working in the fields like they're a 30-year-old," Rangel said. "If you do it the proper way, you can avoid injuries for years."

Rangel said many of the workers who travel to local farms do not come with a background in farming. She knows workers who are nurses, teachers, construction labors and even one who is a judge back home in Mexico. Many are not prepared for physically demanding farm work, whether cutting cabbage, harvesting cucumbers or picking fruit.

Many of the workers are teenagers or in their early 20s. They are away from home for the first time and many have trouble coping with the separation and adjusting to a new culture and the demands of their jobs. Oak Orchard found treatment for 80 workers with "emotional disorders," people typically struggling with homesickness, Linhart said.

"It's like our college students who are away from home," she said. "They're homesick, they're not eating well."

Rangel said the farmworkers have other stresses in their lives. When they arrive locally they often don't have money to spend for at least a week. Oak Orchard collects clothes and blankets to help workers that first week. Even when they have money, arranging trips to the grocery store and Laundromat can be difficult. And many workers struggle adjusting to the cold weather in the fall.

"Everything is stressful in their lives," Rangel said.

Other health conditions

The workers bring pre-existing conditions with them, including 269 workers treated with gastro-intestinal disorders, often from parasites picked up in their home country. Oak Orchard will give them medication to battle those bugs.

Another 518 workers were diagnosed and treated with cardiovascular issues. Oak Orchard counsels them on low-salt diets and offers prescriptions to help reduce the chances for stroke and heart attacks. Another 274 workers were treated for hypertension and 232 for diabetes. She said the workers, many with lean and muscular bodies, may not appear candidates for cardiovascular problems and diabetes.

"They have good cardio fitness, but their circulation could be bad," she said.

Oak Orchard also treated 361 workers who had skin diseases, which could be fungal infections or sunburns. The long exposure to the sun also causes premature cataracts in some workers, Linhart said.

"They get too much ultraviolet exposure and they should have protective glasses," she said.

The steady exposure to loud farm equipment also wears down workers' hearing. Oak Orchard diagnosed hearing problems in 25 workers.

Restivo said she notices many of the workers drop weight and add muscle during their months with local farms.

"These guys are as hard as nails," she said. "Their jobs are a very physical thing and they become very muscular."

When she visits them in the camps, the workers often insist she take a can of soda, and bags of onions, tomatoes and apples.

"They're appreciative," Restivo said. "They always want to give me something. There's such a hospitality and appreciation for our services."

Acknowledgements

I want to thank the farms that hosted me in my odyssey as a "hired hand." The owners of the farms took a great risk having a reporter try some of these physically punishing jobs. I could have been injured and that would have become a paperwork nightmare. But the farms and *The Daily News* wouldn't let concerns about my well-being stop this project. They looked liability in the face and didn't blink – and I hope they said a little prayer I would keep my fingers intact.

It's a powerful experience to have someone take a chance on you, especially when there are many reasons to say no. But the farmers showed faith in me. They told me I could do it. I am grateful for the opportunity and for that support.

I want to also thank my mother Cathy Rivers and late grandmother Reva Pastor. They both had gardens when I was a kid and they often dispatched me to yank the weeds and water the plants. I don't consider myself a farmer, but those gardening efforts as a kid made the farm work seem a little less foreign for a guy who typically types away at a keyboard in an office with central air.

My father Bill instilled a work ethic in me when I was a young kid. He never had qualms about finding jobs for me. I raked leaves, shoveled driveways, cleaned rugs and delivered newspapers as a boy. My father taught me there is honor in every job.

My wife Marsha and I share the household chores and other responsibilities associated with raising three kids. I was less available in 2008 with those all-day farm efforts and the recovery days afterward when my arms didn't work. Then I decided to train for a marathon, which meant less time at home and an exhausted body when I was there. Marsha picked up the slack with no complaints. She also thought of the title for this book.

I fully expected to be inundated with hate mail from *Daily News* readers as these stories trickled out from May to October 2008. When I had written about immigration before, I could count on a few emails from readers saying the Mexicans should be sent home. Some readers thought the farmers should be ashamed for bringing "these people" into our community. I wasn't sure how readers would react to long stories about "these people" and big photos of them doing jobs that some people insisted were being stolen from local Americans.

I bet I received nearly 100 emails or hand-written letters. Only one was negative. The other letter writers thanked me for taking them into the fields and meeting some of the workers who, despite being so visible to passing motorists, are anonymous. The most effusive in their praise tended to be farmers and their wives, especially from an older generation. Those writers, many Polish or Italian immigrants in their 80s and 90s, saw a lot of themselves in the farmworkers, people willing to do any job to give themselves and their families a chance. I treasure those letters and the many handshakes and hugs from readers.

When I pitched this project to the managers at *The Daily News* in March 2008, it didn't sound like a ratings smash. Would people read lengthy first-person accounts of picking fruit and vegetables? I had a feeling they would, but management could have killed the project before it even started. The editors found space for the stories and helped lighten my load by giving someone else the political beat – during a historic election – so I could devote the necessary energy and attention to the farm labor project. I enjoy the freedom at the newspaper. It's a place where you can color outside the lines.

I also need to thank *Daily News* photographers Rocco Laurienzo, Mark Gutman and Nick Serrata for getting out in the fields and the cow barns for many of these photos. Nick ruined a good pair of shoes getting those cabbage pictures. Mark wore a wet suit when he came to the dairy farm. The photographers all immersed themselves on the farms. I appreciate their efforts.

I intended these articles to give readers a glimpse of farm work and to try to find out why so few locals take the jobs. As I described my strained tendons and aching back – symptoms I think we all would expect from hauling enormous pumpkins, stooping to grab cucumbers and lugging 40-pound bushels of apples for hours on end – I hoped readers could relate to the farmworkers through me.

I also hoped that the lack of a common language wouldn't prevent me from communicating effectively with "these people." I found the workers warmed up to me as the day wore on. Some of them knew some English, and I tapped my long-neglected supply of Spanish words and phrases I hadn't used since learning the language un poquito (a wee bit) in high school.

Mostly we let our actions do the talking. I knew I could surprise the workers by merely staying on the job. If I left after an hour or two, I would be just like the other Americans who occasionally try the field work and quickly crumble. When I wouldn't budge from my cucumber row and refused to rub my back from the pain of it, the workers accepted me. Some of the workers seemed to make it their mission to help me survive the full day's work. Their encouragement and support helped sustain me to the end.

I don't want to depict myself as some ultimate warrior or Rambo character. I'm just an average person who summoned every bit of power I could from my rather modest muscles.

Juan Morales, left,of Hulberton and Steve Coville of Albion are pictured in 2002 in front of a school they were helping to build in Afghanistan. (Photo courtesy of Steve Coville)

When I was out in the fields I often thought about two guys I know in Orleans County, two guys in their 40s. Juan Morales and Steve Coville have both served multiple tours in Iraq and Afghanistan, lugging 80 pounds of equipment on their backs in 120-degree heat. I thought of them and how hard they have been working. They and the other American soldiers want to win "hearts and minds." They look for a chance to leave a favorable impression of our country on Iraqis and Afghanis.

I realized I was representing the United States of America when I was with the crews from Mexico, Jamaica and Haiti. Like it or not, these workers would make a judgment about the USA based on my showing that day. Most Americans didn't leave the foreign workers with favorable impressions because the Americans gave up so easily or didn't try hard. The foreign workers figured I was the next hapless American, naive with a knife and weak with willpower.

I knew I had been given a chance to represent my country, to win their hearts and minds. I knew if I stayed and did a good job, these guys might think a little more highly of the American work ethic. It's a chance we all should want, to represent our country, to have the chance to dig deep and find what we're capable of. I'm grateful for that opportunity. I didn't want to blow it.

About the Author

Tom Rivers has worked for *The Daily News* in Batavia, N.Y., since 1997. He covers agriculture, local government and other community stories. Rivers has won several Associated Press writing awards, and was named a "Friend of Agriculture" in 2006 by the Orleans County Farm Bureau. The New York State Agricultural Society presented him with an honorary "golden pitchfork" in January 2009, in appreciation for the farm labor series in *The Daily News.* Rivers also was named a finalist for the Mike Berger Award, given by the Columbia Journalism School for the best in-depth, human-interest reporting. He lives in Albion with his wife Marsha and their four children.